HISTORY SMASHERS

THE *TITANIC*

THE HISTORY SMASHERS SERIES

The Mayflower

Women's Right to Vote

Pearl Harbor

The Titanic

HISTORY SMASHERS

THE *TITANIC*

KATE MESSNER

ILLUSTRATED BY MATT AYTCH TAYLOR

RANDOM HOUSE 🏠 NEW YORK

Library of Congress Cataloging-in-Publication Data
Names: Messner, Kate, author. | Taylor, Matt Aytch, illustrator.
Title: The Titanic / Kate Messner; illustrated by Matt Aytch Taylor.
Description: First edition. | New York: Random House, [2021] | Series: History smashers | Includes bibliographical references and index.
Identifiers: LCCN 2020037630 | ISBN 978-0-593-12043-9 (trade) | ISBN 978-0-593-12044-6 (lib. bdg.) | ISBN 978-0-593-12045-3 (ebook)
Subjects: LCSH: Titanic (Steamship)—History—Juvenile literature. | Shipwrecks—North Atlantic Ocean—History—Juvenile literature.
Classification: LCC G530.T6 M475 2021 | DDC 910.9163/4—dc23

Printed in the United States of America
10 9 8 7 6 5 4
First Edition

For the readers of
Peru Intermediate School

CONTENTS

You've probably heard of the *Titanic*, the great "unsinkable" ship that crashed into an iceberg and sank in the icy North Atlantic, killing more than 1,500 people. Maybe you've read about brave heroes who gave their lives in the service of others on that frigid night, or cowardly villains who only cared about saving themselves. Perhaps you've heard heartbreaking details: how there weren't enough lifeboats and how third-class passengers were locked away from the boat deck, where they might have been saved.

The tale of the *Titanic* is legendary—but only parts of that tragic story are true. When we take a closer look—through eyewitness accounts, historical evidence, and the work of modern-day archaeologists—other parts come crashing down. Here's the real deal about that not-so-unsinkable ship that captured the world's attention after it hit an iceberg on the night of April 14, 1912.

ONE
UNSINKABLE? THE BUILDING OF THE *TITANIC*

Mention the *Titanic,* and most people think of the disaster at sea. But the truth is, the first victims of the *Titanic* died while the ship was still being built.

The shipyard in Belfast, in what is now Northern Ireland, was a dangerous place. Piecing together a nearly nine-hundred-foot-long ship that weighed more than forty-six thousand tons was no small job. With thousands of men working at once, accidents were common.

Sometimes workers dropped tools or rivets—the

Workers leave the Harland and Wolff shipyard
at the end of a shift, 1911.

heavy steel pins that held the ship together. Staging, or scaffolding, collapsed, and people fell. "He's away to the other yard," the men would say when they had to share the sad news of a worker who had died on the job.

The first victim of the *Titanic* was an Irish teenager who fell from a ladder and fractured his skull on April 20, 1910. He'd been part of a riveting crew on the ship. On June 17, 1911, the *Belfast News-Letter* reported that forty-nine-year-old Robert Murphy fell to his death when some staging collapsed. His son, also a *Titanic* shipyard worker, had died in an accident just six months earlier. The following March, the same newspaper carried the story of a man who'd suffered severe injuries "working on a crane when he was crushed in the machinery." There were 254 official accidents recorded during the building of the *Titanic,* including at least eight fatalities.

Believe it or not, that wasn't a bad safety record for a shipyard. The Harland and Wolff shipyard in Belfast, where the *Titanic* was built, was one of the most modern in the world at that time and had a reputation for designing and building high-quality ships.

Harland and Wolff built ships for a company called the White Star Line, which transported passengers and cargo across the North Atlantic, between England and New York. In the early 1900s, two new ships built for that route, the *Titanic* and her sister ship, the *Olympic,* were designed at the Harland and Wolff drawing offices. The architects drew up plans in big rooms full of long tables and natural light. The goal was to lure wealthy passengers from other shipping companies to the White Star Line with a promise of luxury instead of speed.

The shipbuilding project launched in 1907. Because the *Titanic* and the *Olympic* were so big, two new slipways had to be built at the Belfast shipyard before construction of the ships could even start. The project required new machinery, too—a two-hundred-ton floating crane to lift engines, boilers, and funnels for the great ships. Then workers constructed a steel gantry, an enormous structure with cranes, elevators, and walkways. This scaffolding from which workers would build the *Titanic* and her sister ship could be seen from all over Belfast.

The building of the *Titanic* began with the laying

of the ship's keel in March 1909. That's the backbone of the ship; the rest of the vessel would be built around it. Next came the frames, sticking up like ribs, and, eventually, steel plates to cover the ship's skeleton. These were held together by rivets, and workers pounded in three million of them before the job was done.

If you were a young shipyard worker, you might have been part of one of these riveting teams. They were usually made up of three or four men and a boy. The young person's job was to heat a rivet until it was red-hot, take it from the furnace with pincers, and rush it to the area where the other riveters were working. Then the boy—or sometimes another worker, called the holder-on—would put the heated rivet into the hole where the steel plates overlapped. Two men would work outside the hull with heavy hammers to pound the rivet into place. They'd take turns swinging so the pounding never let up. The third man was inside the hull, using an even heavier hammer to hold the rivet in place. With all that hammering, it was so noisy that workers couldn't even hear themselves talk. The shipyard's Belfast neighbors could always hear the bang and rattle of ships coming to life—clangs and creaks and the clash of steel on steel.

It took just over two years to build the *Titanic*'s hull, the main body of the ship. The work went quickly—so quickly that there were rumors a worker had somehow been trapped inside the hull.

Some even said they heard tapping coming from inside, but there's no evidence that this rumor is true. It's possible that what they heard was an inspector tapping with a hammer to test the rivets at the end of the day. But that story made some workers feel certain that something was going to go wrong with the *Titanic*.

Another rumor that fed that feeling had to do with the shipbuilder's hull number, which was 390904. Someone apparently saw that number reflected in

a mirror or in the water and thought it looked like the words "NO POPE." Most people in Belfast were Catholic, so "NO POPE" seemed like a bad phrase to have associated with a new ship. Some shipbuilders decided it meant the *Titanic* was doomed.

But the *Titanic's* construction continued. On May 31, 1911, it was time for the great ship to launch.

THE SHIPYARD BUZZED WITH ACTIVITY ON THE DAY OF THE LAUNCH.

WHITE STAR LINE CHAIRMAN BRUCE ISMAY WAS THERE, ALONG WITH CELEBRITY GUESTS LIKE AMERICAN MILLIONAIRE JOHN PIERPONT MORGAN, WHOSE COMPANY OWNED THE WHITE STAR LINE.

THE SHIPYARD FIRED OFF ROCKETS TO MARK THE START OF THE LAUNCH.

WORKERS KNOCKED LOOSE THE TIMBERS THAT WERE HOLDING THE TITANIC IN PLACE.

THERE SHE GOES!

A SPECIAL APPARATUS HAD BEEN BUILT TO GET THE SHIP STARTED MOVING DOWN THE WAYS, OR THE RAMP LEADING TO THE WATER.

WORKERS HAD GREASED THE WAYS WITH TONS OF SOAP AND TALLOW, OR ANIMAL FAT, TO HELP THE SHIP SLIDE INTO THE WATER.

THE CITY OF BELFAST COULDN'T HAVE BEEN PROUDER.

9

After the launch, the *Titanic* was towed to a deep-water wharf for the second phase of work. Now the ship's interior had to be built. That work took more than three thousand men another ten months to complete. They installed the engines and boilers. They built the passenger cabins, installed all the plumbing and electricity, and painted the hull. When the ship was finally complete, it was something to brag about. And that's exactly what the White Star Line did.

"THE STAIRCASE IS ONE OF THE PRINCIPAL FEATURES OF THE SHIP, AND WILL BE GREATLY ADMIRED AS BEING WITHOUT DOUBT THE FINEST PIECE OF WORKMANSHIP OF ITS KIND AFLOAT."
—WHITE STAR LINE PUBLICITY MATERIALS

The *Titanic* was a floating hotel. In addition to all the cabins and dining rooms, the ship had its own swimming pool, squash court, and gym. There were cafés, palm courts, a barbershop, a dark room for photographers, and a lending library, in case passengers wanted something to read on the voyage.

First-class accommodations were the fanciest, of course, but the *Titanic*'s third-class cabins were also nicer than usual on ships of that time. The White Star Line made it a point to talk about that when promoting its ship to immigrants who were crossing the ocean to settle in America, promising that the third-class general room, where people gathered to visit and play music, would be "one of the liveliest rooms on the ship."

"IN THESE VESSELS THE INTERVAL BETWEEN THE OLD LIFE AND THE NEW IS SPENT UNDER THE HAPPIEST POSSIBLE CONDITIONS."
—WHITE STAR LINE PUBLICITY MATERIALS

Twenty-nine coal-fired boilers supplied steam to the massive ship's engines. The *Titanic* had 159 furnaces and four funnels, or smokestacks. The ship's passengers didn't realize it, but only three of those funnels were actually carrying smoke. The fourth was just for show, to make the ship look more powerful and balanced. When all the work had been completed, the *Titanic* was a sight to behold.

SISTER SHIPS

The *Titanic* might be the most famous ship in history, but if it hadn't hit that iceberg in the North Atlantic, you might not even know its name. That's because the *Titanic* was just one of three big ships the White Star Line planned to build to compete with the Cunard Line, another shipping company that had already launched luxury ocean liners called the *Lusitania* and the *Mauretania*.

The *Titanic* and the *Olympic* were almost identical. They were built side by side at the shipyard in Belfast. But the *Olympic* launched first, on October 20, 1910, and got most of the attention at the time. The *Olympic* sailed from Southampton, England, to New York in June 1911, the same route the *Titanic* would take the following year. It had the same captain and the same shortage of lifeboats, but it

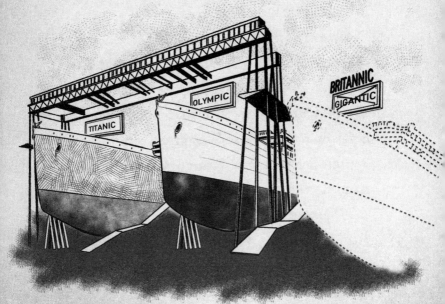

didn't hit an iceberg, so it's the *Titanic* that we remember today.

The third ship the White Star Line had planned was going to be called the *Gigantic,* according to newspaper reports, but the name was changed to the *Britannic* later on. Work on that ship was delayed, but it finally launched in 1914. Before the ship even had a chance to carry passengers, though, World War I began. The *Britannic* was put into service as a hospital ship and ended up hitting a mine and sinking near Greece.

The *Titanic*'s maiden voyage had been scheduled for March 20, 1912, but that plan changed when its sister ship had an accident. The *Olympic* had been damaged when a British navy cruiser rammed into it in September 1911. The *Titanic* had to be moved from its work site so that crews could repair the *Olympic.* When the work was all done, the *Titanic*'s maiden voyage was rescheduled for April 10.

UNSINKABLE?

THAT'S NOT QUITE WHAT WE SAID. . . .

The idea that the *Titanic* was "unsinkable" is part of the mythology surrounding the shipwreck. In a famous movie about the disaster, a character looks up at the ship before it sails and says, "So this is the ship they say is unsinkable." But really, most people weren't talking about that at the time. They were more interested in the ship's palm courts and fancy dining rooms.

The truth is, no one had promised that the *Titanic* was unsinkable. The White Star Line had advertised the *Titanic* as the "largest and finest" ship. One of its brochures for the *Titanic* and the *Olympic* did say the ships were "designed to be unsinkable." Newspapers picked up on this.

BELFAST NEWS-LETTER
JANUARY 1, 1912

Each watertight door can be released by means of a powerful electric magnet controlled from the captain's bridge, so that, in the event of accident, the movement of a switch instantly closes each door, practically making the vessel unsinkable.

After the *Titanic* sank, that "unsinkable" line in the White Star Line publicity materials took on a heavier weight, and everyone was talking about it. It made for a much more dramatic story when an "unsinkable" ship went down.

TWO
SETTING SAIL

Before the *Titanic* could sail across the Atlantic, the ship had to undergo sea trials, or tests in the water to make sure everything was working well. That was scheduled for April 1 but was delayed a day because of wind. Crews were putting the finishing touches on work until the very last minute.

On April 2, the *Titanic* completed her sea trials and left Belfast for the last time, setting sail for Southampton. There, the ship would pick up supplies for the trip. A *lot* of supplies. The journey across the Atlantic with more than two thousand people required literally tons of food.

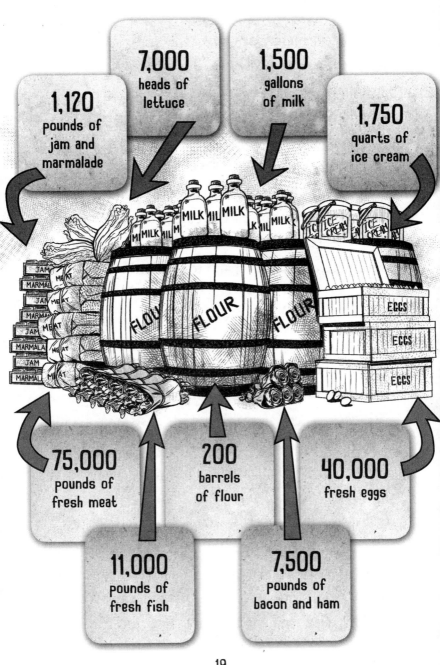

1,120 pounds of jam and marmalade

7,000 heads of lettuce

1,500 gallons of milk

1,750 quarts of ice cream

75,000 pounds of fresh meat

200 barrels of flour

40,000 fresh eggs

11,000 pounds of fresh fish

7,500 pounds of bacon and ham

Fancy dinners planned for the first-class passengers required not only food but fine china and *very* specific silverware. In addition to thousands of cups and plates, workers loaded dinner forks, fruit forks, fish forks, oyster forks, dinner spoons, dessert spoons, egg spoons, teaspoons, salt spoons, and mustard spoons onto the ship. There were towels, blankets, and sheets for all the passengers. The *Titanic* also needed fuel for its voyage, so crews loaded another 4,427 tons of coal into the bunkers.

Most of the *Titanic*'s crew boarded the ship in Southampton—sailors, stewards, deckhands, and stokers, who kept the fires going. It took a lot of staff to feed all those passengers, so the *Titanic*'s crew also included waiters, butchers, bakers, a pastry cook, a fish cook, a "coffee man" and his assistant, a wine butler, plate washers, and other kitchen workers. There were wireless operators to send and receive messages, along with night watchmen and lookouts to watch for icebergs. Every part of the ship seemed to have its own staff. The fitness area had gym instructors, a squash court attendant, and masseuses. There were clothes pressers, a post office clerk, a telephone operator, and an orchestra leader and band members. A ship's bugler was on board to announce meal times, and some doctors came along in case anyone got sick.

A handful of shipyard workers from Belfast had joined the *Titanic*'s maiden voyage as well, in case there was last-minute work to do or small repairs to make during the trip. Shipbuilder Thomas Andrews was on board, too, to oversee any changes or adjustments to the ship that might need to happen on the way.

WHAT DID THEY GET PAID?

The men and women who staffed the *Titanic* earned a wide range of salaries for their work. Here's a sampling of what some workers earned in British currency.

Captain Edward J. Smith
£105/month

Radioman Harold Bride
£48/month

Able Seaman Edward Buley
£5/month

Stewardess
Annie Robinson
£3.10/month

WHAT DID THEY PAY?

Passenger fares were based on the kind of tickets they bought.

3rd class—about £7
More than 700 passengers bought third-class tickets.

2nd class—about £12
About 285 people bought second-class tickets.

1st class—as much as £870
About 325 passengers had first-class tickets that
cost various amounts, depending on the cabin.

"It's like a floating town," passenger Charlotte Collyer wrote to her parents after the *Titanic* left Southampton. "You would not imagine you were on a ship. There is hardly any motion [and] she is so large we have not felt sick. . . ."

Titanic officer Charles Lightoller said the ship was so big that it took him two weeks to find his way around. "It is difficult to convey any idea of the size of a ship like the *Titanic*," he said, "when you could actually walk miles along decks and passages, covering different ground all the time."

From Southampton, the *Titanic* headed for Cherbourg, France, where some people who'd only bought tickets for a cross-channel trip got off the ship. More passengers boarded the *Titanic* in Cherbourg as well. They came from a wide range of countries and backgrounds. While some famous movies about the *Titanic* depict all of the passengers as white, that wasn't the real deal. It's true that there were many English and Irish people on board, but dozens of Syrian passengers boarded in Cherbourg.

So did second-class passenger Joseph Laroche, a

Black Haitian engineer who was leaving France with his family because racism was preventing him from getting a good job there. Laroche died in the wreck, but his wife and two daughters were rescued in a lifeboat.

FURRED AND FEATHERED PASSENGERS

Not all of the *Titanic*'s passengers were human. Pets cost the same as children—half of the adult fare—and there were reportedly twelve dogs on board when the ship set out on its voyage. Philadelphia banker Robert W. Daniel brought along the champion French bulldog he'd just purchased. American millionaire John Jacob Astor had his Airedale named Kitty, and Henry Sleeper Harper brought his Pekingese, a fluffy little dog he'd named Sun Yat-sen, after the president of China. Other canine passengers reported to be on board were a Pomeranian, a King Charles spaniel, a chow chow, a Great Dane, and a little lapdog named Frou-Frou. Three of

the small dogs were rescued in lifeboats with their humans, but the rest went down with the ship. Other animals on the *Titanic*'s maiden voyage included some roosters and hens, a yellow canary, and the ship's cat, named Jenny.

After Cherbourg, the *Titanic* returned to Ireland for one last stop in Queenstown, where seven people got off the ship and another 120 passengers boarded. And then it was time to set off across the Atlantic!

The *Titanic* pulled out of Queenstown on April 11 with 2,201 passengers and crew on board. The first-class passengers sat down to a big breakfast that included ham, sausage, bacon, eggs, rolls, baked apples and fruit, and buckwheat cakes with black-currant jam and honey. The voyage across the Atlantic was officially underway! For three days at sea, the *Titanic*'s passengers would enjoy the fancy dining rooms, the brisk air of the promenades where they walked, and the company of their fellow passengers before their ship hit the iceberg.

The *Titanic*'s first-class dining room

EARLY WARNINGS OF DISASTER?

A handful of *Titanic* passengers reported having an uneasy feeling about the ship's maiden voyage, like something bad might happen. Chief Officer Henry Wilde was supposed to have been given command of a different ship, the *Cymric,* and was disappointed that he ended up being assigned to the *Titanic* instead. Just days before the voyage, he wrote a letter to his sister.

I still don't like this ship. . . . I have a queer feeling about it.

After the disaster, some people pointed to an 1898 novel as another early warning.

Novelist Morgan Robertson wasn't talking about the *Titanic* when he wrote those words. He was talking about the *Titan*, the fictional ship in his 1898 novel *Futility*. The coincidences don't end with the similar ship names. Robertson used the words "unsinkable" and "indestructible" to describe the *Titan*, which hit an iceberg in the story, killing nearly all three thousand people on board. Had he foretold the disaster? Probably not. But after the *Titanic* sank, people rediscovered Robertson's book and pointed out the eerie similarities.

THEY SKIPPED
THE TRIP

Not everyone with plans to sail on the *Titanic*'s maiden voyage ended up making the trip. Ship fireman Thomas Hart was supposed to be on board. In fact, he was among those listed as lost when the *Titanic* went down. The ship's records show that he was on the 8:00 p.m.–12:00 a.m. watch the night the ship hit the iceberg. Hart's mother was devastated . . . until Thomas showed up at home on May 8, alive and well. Turns out he never made it onto the ship.

Hart was all set to go but spent the night before he was supposed to leave in a pub and somehow lost his discharge book, which gave him permission to board. Apparently, it was found (or stolen?), since *somebody* used it

to board the ship and was assigned to that watch. But Hart had no idea who might have ended up taking his place on the crew. He'd been afraid to tell the story about losing his discharge book, so he wandered around for a while before he went home. Even now, no one knows the identity of the person who boarded the ship—and died in the wreck—in Hart's place.

Hart was one of around fifty-five people who had tickets to be on the *Titanic* but didn't make it. Several people canceled when they found out they couldn't get the kind of cabin they wanted. Henry Clay Frick didn't go because his wife sprained her ankle. John Pierpont Morgan was going to use Frick's ticket, but then he couldn't go, either, because a business meeting popped up. Mr. and Mrs. E. W. Bill called off their trip after she had a dream about a shipwreck.

George W. Vanderbilt and his wife had tickets, too, but canceled when George's mom told him that maiden voyages of anything were a bad idea. "So uncomfortable," she said. "So much can go wrong." So the Vanderbilts didn't go. Their luggage went, though, along with a servant named Frederick Wheeler, who was sent to tend the bags and died in the disaster.

WHO'S WHO: A *TITANIC* YEARBOOK

Here are some of the most famous players in the story of the *Titanic*.

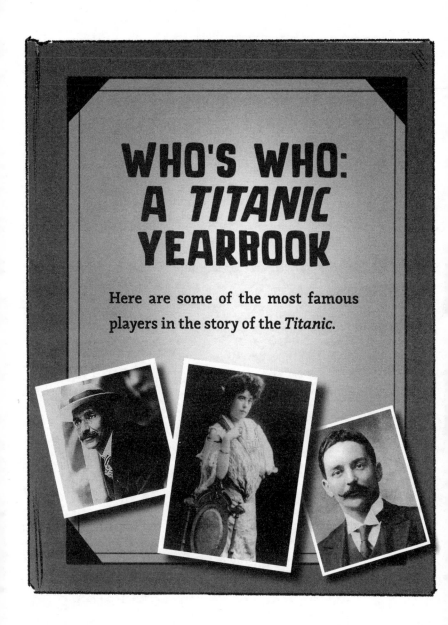

THOMAS ANDREWS was the Harland and Wolff shipbuilder who made the plans for the *Titanic* and the *Olympic*. He was on the maiden voyage and went down with the ship he'd designed.

First-class passenger **JOHN JACOB ASTOR** was an American millionaire—one of the richest men in the world at the time. His fortune was worth an estimated $90 million to $150 million in 1912. (In today's dollars, that would make him a billionaire!) Astor didn't survive the disaster. Workers found him with more than $2,000 when his body was recovered.

JOSEPH BOXHALL was an officer on board the *Titanic*. At the request of the captain, he figured out the ship's exact position in the North Atlantic so that information could be shared in the distress call. Boxhall was put in charge of a life-

boat and survived, along with the passengers he saved.

HAROLD BRIDE was the *Titanic's* junior wireless operator. He helped relay messages after the ship hit the iceberg and was among the last people to escape in a lifeboat.

MOLLY BROWN was a *Titanic* passenger who urged other frightened people to get into the lifeboats. The sailors in her boat were having trouble rowing, so she grabbed an oar and rallied the other women to help, too.

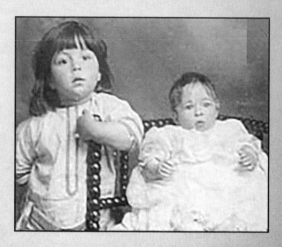

At nine weeks old, **MILLVINA DEAN** was the youngest passenger on board the *Titanic*. Her family was moving to America so her father could open a shop in Kansas. He died in the disaster, but Millvina survived, along with her mother and brother.

FREDERICK FLEET was the *Titanic* lookout who rang the first alarm when the iceberg was spotted. He escaped in a lifeboat.

ROBERT HICHENS was the quartermaster at the wheel when the *Titanic* hit the iceberg. He'd tried to steer the ship away, but the warning had come too late, so the iceberg scraped the side of the ship. Hichens was put in charge of a lifeboat and survived.

BRUCE ISMAY was chairman of the White Star Line. He escaped from the sinking ship in one of the last lifeboats.

NOËL LESLIE, the Countess of Rothes, was another famous *Titanic* passenger. She was traveling with her cousin Gladys Cherry. The two women escaped in a lifeboat, helped the crew row, and reportedly tried to go back to look for survivors but were overruled by other passengers. Later, after Leslie helped care for survivors on the rescue ship, London's *Daily Sketch* newspaper reported that crew members were calling her "the plucky little countess." (Men didn't always speak with much respect for women who took charge of things back then.)

CHARLES LIGHTOLLER was a *Titanic* officer who launched lifeboats after the collision. He was the senior surviving officer, so he ended up being the ship's spokesperson after the wreck.

STANLEY LORD was captain of the *Californian,* a nearby ship that was trapped by ice. Some on the ship might have witnessed the *Titanic* sinking. After the wreck, Lord was criticized for not coming to the *Titanic*'s aid more quickly.

LORD JOHN CHARLES BIGHAM MERSEY was the head of the British inquiry into the *Titanic* disaster.

JAMES MOODY was an officer on duty on the bridge when the *Titanic* hit the iceberg. He died when the ship went down.

WILLIAM M. MURDOCH was first officer in charge the night of the wreck. He tried—too late—to keep the ship from hitting the iceberg and did not survive.

JOHN "JACK" PHILLIPS was the *Titanic*'s senior wireless operator. He spent the night of the disaster trying to find another ship to help and died when the *Titanic* went down.

HERBERT PITMAN was a third officer on board the *Titanic* who helped launch lifeboats after the collision. He was put in charge of a lifeboat and survived.

CAPTAIN ARTHUR ROSTRON was captain of the *Carpathia,* the ship that rescued the *Titanic's* survivors.

EDWARD J. SMITH was captain of the *Titanic*. He went down with his ship.

SENATOR WILLIAM ALDEN SMITH (no relation to Captain Smith) was the head of the US Senate investigation into the *Titanic* disaster.

ISIDOR and **IDA STRAUS** were passengers on the *Titanic* who didn't survive the wreck. Isidor wouldn't board a lifeboat until all the women and children on the ship were safe, and Ida refused to go without her husband.

THREE
STEAMING TOWARD DISASTER

When we talk about the *Titanic,* one of the stories that's told over and over is that no one saw the iceberg until it was too late. That's true, but there were plenty of early warnings that should have let the ship's crew know what they were sailing into.

The winter of 1912 had been mild, so big chunks of ice had broken off and drifted south into the shipping lanes. Ships arriving in New York in early April had reported more ice than usual. The captain of the *Carmania,* which crossed the Atlantic just before the *Titanic,* said he'd never seen an ice field so far south.

And a French ship called the *Niagara* had just crashed into one of those icebergs—a head-on collision that threw passengers from their chairs and sent dishes crashing to the floor. But everyone survived, and damage to the ship was minor enough that the *Niagara* still made it to port.

The *Titanic* was entering that same icy area of the North Atlantic on April 14, 1912. It was a Sunday, so the morning began with a church service, complete with music.

OH GOD, OUR HELP IN AGES PAST . . .

Even as the passengers were singing hymns, there were warnings of ice ahead. At 9:00 a.m., the *Titanic* received a wireless message from another ship, the *Caronia,* which was on its way from New York to Liverpool, England.

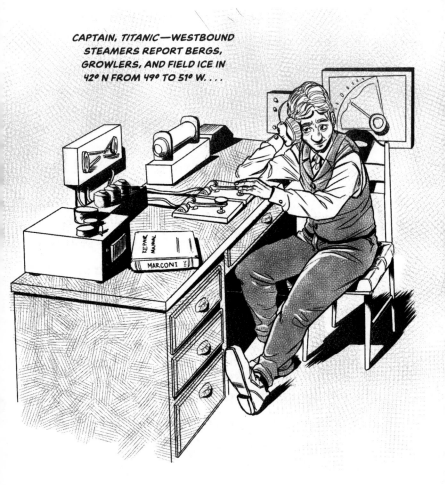

CAPTAIN, TITANIC—WESTBOUND
STEAMERS REPORT BERGS,
GROWLERS, AND FIELD ICE IN
42° N FROM 49° TO 51° W. . . .

That message was delivered to Captain Smith on the *Titanic*'s bridge, and he posted it for his officers to read, too. At 1:42 p.m., another warning came through. The captain of the Greek steamer *Athinai* reported "passing icebergs and large quantities of field ice" within a few miles of the *Titanic*'s path.

You might think that the ship would slow down a bit with those ice warnings coming in, but it didn't. Was that because the White Star Line was trying to get the *Titanic* across the ocean even faster than the *Olympic* had made the journey?

Titanic was scheduled to arrive in New York the morning of Wednesday, April 17. But several survivors reported that Bruce Ismay had talked about arriving early, on Tuesday night. Passenger John B. Thayer said that, before the wreck, Ismay told him that two more boilers were about to be opened up so the *Titanic* could go even faster. Charles Hallace Romaine, a first-class passenger who escaped in a lifeboat, said that people on board were even placing bets on how fast they'd arrive. Survivor Elizabeth Lines testified that she heard Ismay talking with Captain Smith on Saturday, April 13—the day before the *Titanic* hit the iceberg.

I HEARD "WE WILL BEAT THE *OLYMPIC* AND GET INTO NEW YORK ON TUESDAY" IN THOSE WORDS.

Bruce Ismay claimed he never said that. But whatever the reason, the *Titanic* didn't slow down. The captain did order an iceberg watch, though, as passengers were getting ready for dinner.

At 9:40 p.m., there was another message, this time from the *Mesaba*, reporting "much heavy pack ice and great number large icebergs. Also field ice. Weather good, clear."

That warning sounds even more serious, doesn't it? So you're probably thinking that the *Titanic*'s wireless operator, Jack Phillips, rushed right up to Captain Smith on the bridge. He didn't, though. He was too busy.

The *Titanic*'s wireless system hadn't been working correctly earlier that day, so there was a backlog of passenger messages. Phillips was trying to get caught up, so he set the message aside. And that urgent ice warning never made it to the bridge.

Just before 11:00 p.m., a message came from a nearby ship, the *Californian*.

WE ARE STOPPED AND SURROUNDED BY ICE.

The *Californian* was nearby. And surrounded by ice! Surely this would be the warning that got Phillips's attention, right?

Wrong. He literally told the other ship's wireless operator to "shut up." Phillips said he was busy sending passengers' messages, and he didn't want incoming messages jamming things up. So Captain Smith never got that message, either. And the *Titanic* steamed onward.

FREDERICK FLEET AND REGINALD LEE WERE IN THE LOOKOUT CAGE THAT NIGHT, KEEPING WATCH FOR ICEBERGS.

THEY SHOULD HAVE HAD BINOCULARS.

THERE HAD BEEN A PAIR IN THE CAGE WHEN THE SHIP LEFT BELFAST, BUT NOBODY KNEW WHAT HAD HAPPENED TO THEM.

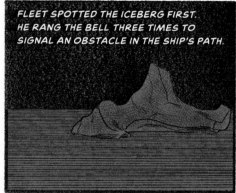

FLEET SPOTTED THE ICEBERG FIRST. HE RANG THE BELL THREE TIMES TO SIGNAL AN OBSTACLE IN THE SHIP'S PATH.

CLANG CLANG CLANG

THEN HE CALLED THE BRIDGE.

ARE YOU THERE?

YES. WHAT DID YOU SEE?

ICEBERG RIGHT AHEAD.

THANK YOU.

FIRST OFFICER WILLIAM MURDOCH TELEGRAPHED THE ENGINE ROOM. STOP: FULL SPEED ASTERN

DING DING

DING DING

HARD-A-STARBOARD!

AND HE CALLED TO THE HELMSMAN.

58

HE PULLED A LEVER TO AUTOMATICALLY CLOSE FIFTEEN WATERTIGHT DOORS IN THE ENGINE AND BOILER ROOM BULKHEADS.

THE *TITANIC'S* HELMSMAN SPUN THE WHEEL.

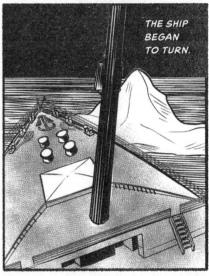

THE SHIP BEGAN TO TURN.

BUT IT WASN'T FAST ENOUGH.

The iceberg was huge. Why hadn't the lookouts spotted it sooner? Aside from the lost binoculars, there were a few reasons. First, the night was calm. You might think that would be a good thing for ships, but it makes icebergs practically invisible. A breeze would have created little waves against the iceberg, making it easier to see.

Also, about four-fifths of an iceberg is under-water. As these giant ice hunks melt, they become lopsided and topple over in the sea, so a different part is above the water, making it wet, dark, and especially hard to spot. That had just happened to the iceberg the *Titanic* hit. To make matters worse, sharp spurs of ice can stick out from the underwater parts, creating an invisible hazard.

Experts believe it may have been one of those spurs that gouged the side of the *Titanic,* opening up half a dozen small gashes in the starboard side of its hull. Those openings were below the water line, so ocean poured into the boiler rooms. The ship's pumps simply couldn't keep up.

When the *Titanic* first hit the iceberg, though, most people weren't terribly concerned. Some felt the

impact, but it didn't seem like a big deal. One passenger called the jolt "slight" and said he'd heard "a jangling as that of chains being dragged along the side of the ship." Another said it felt "as though we went over a thousand marbles." One woman said it was "as though someone had drawn a giant finger all along

Those toward the front of the boat felt the most impact. Third-class passenger Edward Dorking was playing cards with friends when they heard a grinding noise. He said the impact threw them right off their bench.

It was loudest of all in the belly of the ship, where firemen shoveled coal into the boilers.

ROOOAAR

After the iceberg hit, the *Titanic*'s stewards kept polishing brass and doing their usual jobs. They did their best to convince people that everything was fine, even though there were early signs that it wasn't. Shortly after the crash, first-class passengers Norman and Bertha Chambers saw postal workers in a hallway. The postal office was already flooded; their

63

pants were wet up to the knees, and they carried armloads of soggy letters. But a passing steward reassured the couple, saying, "Everything is all right now. . . . You may turn in."

Most people accepted that message. They thought the ship was too big to be impacted by an iceberg. One passenger who went up on deck joked about the ice scattered around after the impact.

I'VE BROUGHT YOU SOME ICE FOR YOUR DRINK.

All in all, passengers weren't concerned. Most were already in their pajamas, cozied up in their rooms for the night. Those who knew about the iceberg collision understood there might be damage. Maybe some scraped paint or a busted propeller. But surely everything would be fine. Most never imagined how quickly things would change—or how serious that scrape with the iceberg had been.

FOUR
URGENT CALLS FOR HELP

It was 11:40 p.m. when the *Titanic* hit the iceberg. Captain Smith rushed up to the bridge.

At first, even Captain Smith thought everything would be fine. At 11:52 p.m., he sent a message to the White Star Line to say that his ship had hit an iceberg but that everyone was safe and he was heading for Halifax, Nova Scotia.

Passengers asked Officer Charles Lightoller if he thought the situation was serious.

"I TRIED TO CHEER THEM UP BY TELLING THEM 'NO,' BUT THAT IT WAS A MATTER OF PRECAUTION TO GET THE BOATS IN THE WATER, READY FOR ANY EMERGENCY. THAT IN ANY CASE THEY ARE PERFECTLY SAFE, AS THERE WAS A SHIP NOT MORE THAN A FEW MILES AWAY, AND I POINTED OUT THE LIGHTS ON THE PORT BOW WHICH THEY COULD SEE AS WELL AS I COULD."
—TITANIC OFFICER CHARLES LIGHTOLLER

Some of the first reports of damage came from those *Titanic* postal clerks, who reported water streaming into the mailroom on the G deck.

Soon after that, it became clear that soggy mail was the least of the crew's worries. If the *Titanic* had hit the iceberg head-on, damage would have been limited to the bow. But because the ship tried to turn to avoid it, the iceberg scraped along the *Titanic*'s side, slicing through its steel plates in multiple places, flooding six of its watertight compartments.

Those safety features had been designed so that the *Titanic* could still float with up to four of the compartments flooded. But not six.

As soon as the ship's builder and captain learned about the damage, they understood what was about to happen. The *Titanic* was going to sink. So they prepared for the worst.

12:05 A.M.

IT WAS ALL HANDS ON DECK.

THE *TITANIC'S* CREW TOOK THE COVERS OFF THE LIFEBOATS TO PREPARE THEM FOR LOWERING . . .

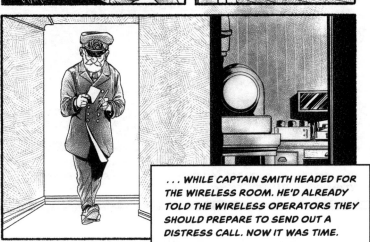

. . . WHILE CAPTAIN SMITH HEADED FOR THE WIRELESS ROOM. HE'D ALREADY TOLD THE WIRELESS OPERATORS THEY SHOULD PREPARE TO SEND OUT A DISTRESS CALL. NOW IT WAS TIME.

WHEN THAT DISTRESS CALL REACHED THE *VIRGINIAN,* THE WIRELESS OPERATOR REPORTED IT TO HIS OFFICER, WHO THOUGHT HE MUST BE PLAYING A PRACTICAL JOKE. NO ONE THOUGHT THE *TITANIC* COULD BE IN TROUBLE.

AS OTHER SHIPS BEGAN TO ANSWER, PHILLIPS TOLD THEM THE *TITANIC* WAS SINKING BY THE HEAD. THE BOW WAS ALREADY STARTING TO GO DOWN.

BUT MOST WERE TOO FAR AWAY TO HELP.

CARPATHIA WIRELESS OPERATOR HAROLD THOMAS COTTAM MISSED THE *TITANIC'S* FIRST DISTRESS CALL BUT CONNECTED WITH PHILLIPS JUST AS HE WAS ABOUT TO GO TO BED.

I SAY, OLD MAN, DO YOU KNOW THERE IS A BATCH OF MESSAGES COMING THROUGH FOR YOU FROM MCC [CAPE COD]?

HE WAS LETTING PHILLIPS KNOW HE'D PICKED UP SOME WIRELESS MESSAGES INTENDED FOR THE *TITANIC.*

COTTAM TOOK OFF HIS HEADPHONES AND RACED FOR THE BRIDGE.

OFFICERS CHECKED THE CHARTS. THE *CARPATHIA* WAS FIFTY-EIGHT MILES AWAY FROM WHERE THE *TITANIC* WAS GOING DOWN.

72

THE *CARPATHIA* HAD LEFT NEW YORK THREE DAYS EARLIER ON A CRUISE TO THE MEDITERRANEAN.

NOW THE SHIP TURNED AND HEADED TOWARD THE *TITANIC*.

OFFICERS TOLD THE CREW TO PREPARE FOR SURVIVORS.

MEANWHILE, CAPTAIN SMITH MET WITH THE *TITANIC'S* DESIGNER, THOMAS ANDREWS . . .

. . . AND LEARNED THAT THEY HAD ABOUT TWO HOURS BEFORE THE SHIP WOULD GO DOWN.

THE *CARPATHIA* WAS COMING, BUT THERE WAS ICE ALL AROUND. THERE WAS NO WAY THE RESCUE SHIP WOULD MAKE IT IN TIME.

Right about now, you might be wondering about that other ship, the *Californian*. Hadn't its wireless operator said it was pretty close to the *Titanic* when he sent that message about being surrounded by ice? And hadn't Lightoller pointed out lights on a nearby ship when he was telling passengers that everything would be all right? Why wasn't that closer ship coming to help?

The truth is, the *Californian* was probably closer to the *Titanic* than the *Carpathia* was. But its wireless operator didn't hear the *Titanic*'s first distress calls. After Phillips responded to the *Californian*'s earlier call with that message to "shut up" because he was busy, the *Californian*'s wireless operator had taken off his headphones and shut down his equipment for the night. Wireless operators didn't monitor their equipment around the clock back then. If they had, the *Californian* would have heard the *Titanic*'s first distress calls and might have been close enough to help.

But now the clock was ticking. The *Titanic* was going to sink in a matter of hours. It was time to start launching lifeboats.

LIFESAVING TECHNOLOGY

The technology that allowed the *Titanic* to send out its distress calls was fairly new at the time. In the 1830s, Samuel Morse had come up with a system of sending telegraph messages using dots, dashes, and spaces to represent letters of the alphabet. Those were sent via electrical pulses of different lengths—short ones for dots and longer ones for dashes. So, for example, one dot would represent *E*. One dash stood for *T*. *A* was dot-dash, and *N* was dash-dot.

Early telegrams had required wires, but by the early 1900s, Italian inventor Guglielmo Marconi had come up with a system for sending those messages wirelessly, via radio waves. American president Theodore Roosevelt tested the technology with England's King Edward VII in 1903.

I EXTEND ON BEHALF OF THE AMERICAN PEOPLE MOST CORDIAL GREETINGS. . . .

I THANK YOU MOST SINCERELY FOR THE KIND MESSAGE WHICH I HAVE JUST RECEIVED FROM YOU THROUGH MARCONI'S TRANSATLANTIC WIRELESS TELEGRAPHY.

BY 1912, THE SYSTEM WAS BEING USED TO SEND SHIP-TO-SHIP AND SHIP-TO-SHORE MESSAGES. WITHOUT IT, IT'S LIKELY THAT NONE OF THE *TITANIC'S* PASSENGERS WOULD HAVE BEEN RESCUED.

FIVE
TO THE
LIFEBOATS!

One of the myths surrounding the story of the *Titanic* disaster has to do with the lifeboats. You've probably heard stories about how the luxury ship cut corners, how there weren't the required number of lifeboats on board, and that's why so many people died. But that's not entirely true.

It *is* true that the *Titanic* only had enough lifeboats for about half of the passengers on board. But even though there weren't enough boats for everyone, the White Star Line actually included *more* lifeboats than the law required. Back then, the number of lifeboats required on a ship was based on the weight of the ship. It had nothing to do with the number of passengers on board who might need to be rescued. At the time, that law said that any ship weighing at least ten thousand tons had to have sixteen lifeboats. The *Titanic* weighed way more than that. It came in at forty-six thousand tons, but there was no difference in how many lifeboats were required, because the law hadn't caught up to the shipbuilding industry, which was now making bigger and bigger ships. So even though it was enormous, the *Titanic* would have been perfectly legal with just sixteen lifeboats on board. But the ship actually had twenty.

THE *TITANIC*'S LIFEBOATS

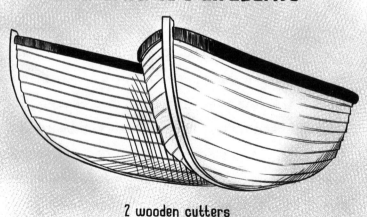

2 wooden cutters
(40 passengers each)

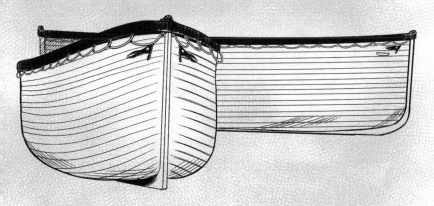

14 regular wooden lifeboats
(65 passengers each)

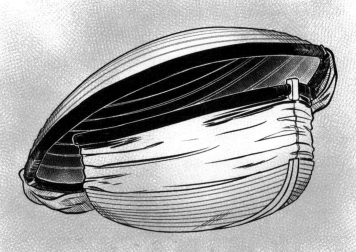

4 collapsible boats (47 passengers each). These had bottoms
made of wood and cork, with canvas sides that could be lifted
before launch or even in the water. They could be stacked on
deck, so they took up less space than regular wooden lifeboats.

Originally, the plan was for the *Titanic* to have more than twice as many lifeboats. Early plans called for forty-eight boats, but the shipbuilders decided that would clutter up the decks and leave passengers less room for walking in the ocean air. Historians also wonder if the White Star Line was worried that more boats could actually make passengers feel *less* safe. After all, if the ship was so modern and safe, why would it need all those lifeboats? So the decision was made to keep just twenty lifeboats—enough to fit less than a third of the ship's maximum capacity. Historians wonder if more lifeboats would have made a big difference. Some argue it wouldn't have, because there simply wouldn't have been time to launch them all. Even with twenty lifeboats, the crew ran out of time to load passengers before the ship went down.

Part of the problem was that people didn't *want* to get into the lifeboats. Half an hour after the *Titanic* hit the iceberg, the first lifeboats were ready for passengers. But there were few passengers to be found. Many were still sleeping. Others knew the ship had hit an iceberg but thought everything was fine. Even after they realized that the crew was loading lifeboats, many weren't interested. It was cold out, and staying on the big ship felt safer than being lowered down in that little lifeboat. Even if the *Titanic* was sinking, another ship was coming to rescue them, right? Why not wait where it was more comfortable?

At about 12:30 a.m., the *Titanic*'s crew was ordered to load the women and children into lifeboats. Crew members walked from cabin to cabin, waking people up. Around ten minutes later, the first boat was lowered.

THE *TITANIC*'S OFFICERS FIRED OFF DISTRESS ROCKETS AS THE EARLY BOATS WERE LAUNCHED. LIFEBOAT 7 WAS THE FIRST TO BE LOADED. IT COULD HAVE HELD SIXTY-FIVE PEOPLE...

LADIES, THIS WAY...

...BUT AT THE TIME, FEW PASSENGERS WANTED TO GO.

MURDOCH ALLOWED MEN TO BOARD, TOO, WHEN HE COULDN'T FIND ENOUGH WOMEN TO FILL THE SEATS.

EVEN SO, LIFEBOAT 7 WAS LOWERED WITH ONLY ABOUT TWENTY-SEVEN PEOPLE ON BOARD.

LATER, THERE WOULD BE RUMORS THAT A MAN NAMED WILLIAM SLOPER SNUCK ONTO THAT BOAT DISGUISED AS A WOMAN.

THAT WASN'T TRUE. SLOPER WAS ON THE BOAT, BUT HE'D BEEN OFFERED A SEAT BECAUSE THE WOMEN DIDN'T WANT TO GO. A TABLOID NEWSPAPER REPORTED THAT FALSE STORY ABOUT HIM, THOUGH, AND HE SPENT THE REST OF HIS LIFE DENYING IT.

FOR THE FIRST HOUR OR SO, THE LIFEBOATS WERE LAUNCHED QUIETLY, FEW OF THEM FULL.

BUT AS TIME PASSED AND THE SHIP BEGAN TO TIP, PEOPLE CHANGED THEIR MINDS. THE DECK WAS IN CHAOS.

1:30 A.M.

OFFICERS HELPED DOZENS OF WOMEN BOARD LIFEBOAT 14. ONE BRANDISHED HIS GUN TO KEEP MEN FROM RUSHING THE BOAT.

SEVEN-YEAR-OLD EVA HART CLIMBED INTO LIFEBOAT 14 WITH HER MOTHER. HER FATHER STAYED BEHIND.

HOLD MUMMY'S HAND AND BE A GOOD GIRL....

OTHER FAMILIES WERE SEPARATED, TOO.

SOMEONE GRABBED A BABY FROM THIRD-CLASS PASSENGER LEAH AKS AND TOSSED HIM INTO A LIFEBOAT.

ANOTHER WOMAN CAUGHT THE CHILD AND WRAPPED HIM IN A STEAMER BLANKET. AKS WAS PUSHED INTO A DIFFERENT BOAT—WITHOUT HER BABY.

WOMEN AND CHILDREN FIRST.

EIGHT-YEAR-OLD MARSHALL DREW WAS WAITING ON THE DECK WITH HIS AUNT AND UNCLE WHEN CREW BEGAN LOADING ANOTHER BOAT.

MARSHALL HAD TO SAY GOODBYE TO HIS UNCLE JIM BEFORE BOARDING THE LIFEBOAT WITH HIS AUNT.

HE COULD HEAR THE SHIP'S ORCHESTRA PLAYING AS THE LIFEBOAT WAS LOWERED INTO THE SEA.

AT FIRST, OFFICERS WOULDN'T LET THIRTEEN-YEAR-OLD JACK RYERSON BOARD LIFEBOAT 4 WITH HIS MOTHER.

BUT EVENTUALLY HE WAS ALLOWED TO GO.

THERE WAS ANOTHER SURVIVOR-IN-DISGUISE STORY TOLD ABOUT RYERSON. RUMOR HAD IT THAT WHEN HE WASN'T ALLOWED TO BOARD, FIRST-CLASS PASSENGER JOHN JACOB ASTOR PUT A WOMAN'S HAT ON HIS HEAD.

THERE, NOW YOU'RE A GIRL AND YOU CAN GO.

BUT THERE'S NO EVIDENCE TO SUPPORT THAT STORY, EITHER. IT WAS NEVER MENTIONED IN ANY OF THE FORMAL REPORTS OR AFFIDAVITS THAT CAME OUT LATER.

PEOPLE GREW MORE AND MORE FRANTIC.

ABOUT 2:00 A.M.

AS COLLAPSIBLE LIFEBOAT C WAS BEING LOWERED, WHITE STAR LINE CHAIRMAN BRUCE ISMAY CLIMBED ABOARD.

HE'D BE CRITICIZED IN THE PRESS FOR LEAVING WHILE WOMEN AND CHILDREN WERE STILL ON THE SHIP.

SHORTLY AFTER 2:00 A.M., THE TITANIC'S BOW WAS UNDERWATER.

JUST TWO COLLAPSIBLE LIFEBOATS REMAINED.

AS CREWS UNTIED THEM, A WAVE WASHED OVER THE DECK.

IT SWEPT THE LAST TWO BOATS OUT TO SEA BEFORE THEY COULD BE LOADED AND LAUNCHED.

Now all twenty lifeboats were gone. If they'd been launched at capacity, they could have saved more than 1,100 people. But most of the boats weren't even close to full. One had left with only a dozen people on board. In the end, only 698 people made it into the lifeboats. The rest of the passengers and crew were trapped on a sinking ship.

Who made it into the lifeboats? More women than men. Because of the "women and children first" policy, three-quarters of the *Titanic*'s female passengers would be rescued, while just one in five men survived. And first-class passengers were more likely to survive than those from lower classes.

FIRST CLASS

39%
died

61%
survived

SECOND CLASS

42% survived

58% died

THIRD CLASS

25% survived

75% died

One of the myths surrounding the *Titanic* disaster has to do with these statistics. There were rumors that third-class passengers were deliberately kept from the lifeboats by locked gates and bullying crew members. But there's no real evidence that ever happened. There were gates because immigration laws at that time required third-class passengers to be kept separate from the rest of the ship to prevent the spread of disease. That didn't really make much sense, since wealthy people get sick, too, but it was the policy back then.

But historians don't believe most of those gates were locked. More likely, other factors kept many third-class passengers from getting into the boats. The third-class cabins were farther from the boat decks. The ship was enormous, with hard-to-navigate hallways, stairs, and passages, so it couldn't have been easy to find their way. Some didn't want to leave behind their possessions; they were traveling with everything they owned. And many of the third-class passengers were immigrants who didn't speak English, the language in which the crew members were giving orders and warnings. It's likely that many simply didn't know what was happening or understand what they were supposed to do until it was too late.

You might think that with so much loss of human life, the dogs on board the ship would be doomed. But three of them were rescued, too—little lapdogs whose owners quietly brought them into the lifeboats.

There were also rumors of a pig in one of the lifeboats. Those were true . . . sort of.

THE GOOD-LUCK PIG

One of *Titanic*'s survivors, Edith Rosenbaum, refused to get into the lifeboat without her good-luck pig. It was a little toy music box that her mother gave her for good luck after she'd survived a car crash that killed her fiancé. Rosenbaum had felt like she needed luck on the trip. She'd written a letter from Queenstown, telling her secretary back home that she liked the ship but "I cannot get over my feeling of depression and premonition of trouble."

When the iceberg hit, Rosenbaum didn't want to get into a lifeboat. But one of the sailors grabbed her musical pig and tossed it into the lifeboat, so she decided to go in after it. It

was her lucky pig, after all. That night, she comforted the children in the lifeboat by making it play its tune.

The pig was rescued along with its owner. Later on, experts at the National Maritime Museum in Greenwich, England, fixed the toy and figured out how it worked. (You could wind the music box by turning its curly tail.) They made a recording of the song and posted it online so the public could help identify the tune. It turned out to be a song called "La Sorella," composed by Charles Borel-Clerc.

SIX

A RACE AGAINST TIME

Those who were left on the *Titanic* could only hope for a miracle. The nearby ship whose lights people had seen never arrived to help. But other ships that had heard the *Titanic*'s distress calls responded.

We are rushing to you.
—*Baltic,* 1:37 a.m.

Am lighting up all possible boilers as fast as can.
— *Olympic,* 1:40 a.m.

Another ship, the *Mount Temple*, also responded to the *Titanic's* distress call; at 12:30 a.m., it was about fifty miles out. The *Carpathia* was on its way, too. Its radio operator tried to stay in contact with the *Titanic,* but communication from the sinking ship ended just after 1:45 a.m. By then the *Titanic's* engine room was flooded up to the boilers.

Members of the *Titanic's* crew were among the last survivors to leave the ship.

THE *TITANIC'S* JUNIOR WIRELESS OPERATOR, HAROLD BRIDE, WAS HELPING WITH THE LAST COLLAPSIBLE LIFEBOATS WHEN HE HEARD CAPTAIN SMITH'S VOICE.

MEN, YOU HAVE DONE YOUR FULL DUTY. YOU CAN DO NO MORE.

ABANDON YOUR CABIN. NOW IT'S EVERY MAN FOR HIMSELF. YOU LOOK OUT FOR YOURSELVES. I RELEASE YOU.

THAT'S THE WAY OF IT AT THIS KIND OF A TIME. EVERY MAN FOR HIMSELF.

THE SHIP WAS TIPPING FORWARD EVEN MORE. BRIDE WAS STILL WORKING TO FREE THE BOAT WHEN THAT BIG WAVE WASHED OVER EVERYTHING.

LATER HE TOLD THE NEW YORK TIMES THE STORY OF HOW HE SURVIVED.

"THE BIG WAVE CARRIED THE BOAT OFF. I HAD HOLD OF AN OARLOCK AND I WENT OFF WITH IT. THE NEXT I KNEW I WAS IN THE BOAT."

"THE BOAT WAS UPSIDE DOWN AND I WAS UNDER IT. AND I REMEMBER REALIZING I WAS WET THROUGH, AND THAT WHATEVER HAPPENED I MUST NOT BREATHE,

FOR I WAS UNDER WATER."

"I KNEW I HAD TO FIGHT FOR IT, AND I DID."

"HOW I GOT OUT FROM UNDER THE BOAT I DO NOT KNOW, BUT I FELT A BREATH OF AIR AT LAST."

THE SEA WAS DOTTED WITH HUNDREDS OF PEOPLE IN LIFE JACKETS . . .

ALL STRUGGLING IN THE FRIGID WATER.

93

BRIDE UNDERSTOOD ONE THING IMMEDIATELY: HE HAD TO GET AWAY FROM THE SINKING SHIP.

"SMOKE AND SPARKS WERE RUSHING OUT OF HER FUNNEL. THERE MUST HAVE BEEN AN EXPLOSION, BUT WE HAD HEARD NONE. WE ONLY SAW THE BIG STREAM OF SPARKS."

"THE SHIP WAS GRADUALLY TURNING ON HER NOSE—JUST LIKE A DUCK DOES THAT GOES DOWN FOR A DIVE. I HAD ONLY ONE THING ON MY MIND— TO GET AWAY FROM THE SUCTION."

"I SWAM WITH ALL MY MIGHT. I SUPPOSE I WAS 150 FEET AWAY WHEN THE *TITANIC*, ON HER NOSE, WITH HER AFTER-QUARTER STICKING STRAIGHT UP IN THE AIR, BEGAN TO SETTLE—SLOWLY."

IT WAS THE SAME COLLAPSIBLE HE'D BEEN WORKING ON WHEN HE WAS SWEPT OFF THE SHIP. HE SETTLED ONTO IT AND SAID A PRAYER.

OUR FATHER, WHO ART IN HEAVEN . . .

BRIDE MADE HIS WAY TO A LIFEBOAT.

94

THE BAND PLAYED ON
(BUT WHAT DID IT PLAY?)

One of the most famous *Titanic* legends is the story of how the band members remained on the deck, playing music as the great ship went down. This is one of the stories that's supported by witness accounts. Multiple survivors reported hearing the band play as they left the ship in lifeboats, and even in the last moments before it sank.

But just what the band was playing depends on who you ask. One passenger said its final song was a hymn called "Nearer, My God, to Thee." That was reported in a newspaper, even though that passenger got off the ship before the band was finished playing. The story was spread in books and movies later on, perhaps because that song's lyrics were so somber and fitting for a nighttime disaster at sea.

Though like the wanderer,
The sun gone down,
Darkness be over me,
My rest a stone,
Yet in my dreams I'd be
Nearer, my God, to thee,
Nearer, my God, to thee,
Nearer to thee!

But that popular story doesn't exactly hold up, because other passengers reported hearing different songs from the lifeboats. *Titanic* officer Charles Lightoller said he heard an upbeat tune as he was loading lifeboat 6. "I could hear the band playing a cheery sort of music," he said. "I don't like jazz music as a rule, but I was glad to hear it that night. I think it helped us all."

Harold Bride said he heard ragtime music and then, after he was swept off the ship with that collapsible lifeboat, a song called "Autumn," which also had comforting lyrics.

Hold me up in mighty waters,
Keep my eyes on things above—
Righteousness, divine atonement,
Peace, and everlasting love.

The truth is, there wasn't just one band on board the *Titanic*. There were two—a saloon orchestra with five musicians and a deck band with three. But it's likely that they all got together to play on the deck at the end. None of them survived. Whatever they played as the final song that night, they are remembered as heroes whose music helped keep passengers calm on a terrible night.

THE ILLUSTRATED LONDON NEWS, April 27, 1912.—638

BRAVE AS THE "BIRKENHEAD" BAND: THE "TITANIC'S" MUSICIAN HEROES.

1. MR. F. CLARKE, OF LIVERPOOL. 2. MR. P. C. TAYLOR, OF CLAPHAM.
3. MR. G. KRINS, OF BRIXTON, SOMETIME OF THE RITZ HOTEL ORCHESTRA. 4. MR. W. HARTLEY (BANDMASTER), OF DEWSBURY. 5. MR. W. T. BRAILEY, OF NOTTING HILL.
6. MR. J. HUME, OF DUMFRIES. 7. MR. J. W. WOODWARD, OF HEADINGTON, OXON.

Once all of the *Titanic*'s lifeboats were gone, those who remained had a decision to make. Should they cling to the rails and try to stay on board as long as possible, or should they jump and hope to be rescued from the sea? Shipbuilder Thomas Andrews remained on the ship. Some survivors said they saw him in the last moments before the ship went down, throwing deck chairs overboard, trying to provide something to float on for those already in the water.

No one knows for sure what happened to Captain Smith, but some survivors reported seeing him jump from the *Titanic*'s bridge into the sea. Second officer Charles Lightoller had thought about trying to make his way to the stern to stay out of the water longer but decided against it.

"IT WOULD ONLY BE POSTPONING THE PLUNGE," HE WROTE, "AND PROLONGING THE AGONY—EVEN LESSENING ONE'S ALREADY SLIM CHANCES, BY BECOMING ONE OF A CROWD. IT CAME HOME TO ME VERY CLEARLY HOW FATAL IT WOULD BE TO GET AMONGST THOSE HUNDREDS AND HUNDREDS OF PEOPLE WHO WOULD SHORTLY BE STRUGGLING FOR THEIR LIVES IN THAT DEADLY COLD WATER."

"THERE WAS ONLY ONE THING TO DO, AND I MIGHT JUST AS WELL DO IT AND GET IT OVER. . . ."

SO HE DOVE FROM THE SHIP'S BRIDGE.

"STRIKING THE WATER WAS LIKE A THOUSAND KNIVES BEING DRIVEN INTO ONE'S BODY, AND, FOR A FEW MOMENTS, I COMPLETELY LOST GRIP OF MYSELF— AND NO WONDER FOR I WAS PERSPIRING FREELY, WHILST THE TEMPERATURE OF THE WATER WAS TWENTY-EIGHT, OR FOUR BELOW FREEZING."

LIGHTOLLER SPOTTED THE SHIP'S LOOKOUT CAGE, STICKING UP FROM THE WATER.

HE STARTED SWIMMING FOR IT . . .

. . . UNTIL HE REALIZED THE SINKING SHIP OFFERED NO SAFETY.

LIGHTOLLER TRIED TO SWIM AWAY BUT FOUND HIMSELF TRAPPED AGAINST THE WIRE NETTING OF AN AIR SHAFT AND VENTILATOR AS WATER POURED IN.

"I WAS STILL STRUGGLING AND FIGHTING WHEN SUDDENLY A TERRIFIC BLAST OF HOT AIR CAME UP THE SHAFT, AND BLEW ME RIGHT AWAY FROM THE AIR SHAFT . . ."

IF THE WIRE GAVE WAY, HE'D BE SUCKED INTO THE DEPTHS OF THE SHIP!

". . . AND UP TO THE SURFACE."

BUT IN AN INSTANT, HE WAS SUCKED UNDERWATER AGAIN, TRAPPED AGAINST ANOTHER GRATE.

SOMEHOW HE MADE IT BACK TO THE SURFACE AND FOUND HIMSELF ALONGSIDE ONE OF THE LIFEBOATS HE'D HELPED LAUNCH.

"THE BOW OF THE SHIP WAS NOW RAPIDLY GOING DOWN AND THE STERN RISING HIGHER AND HIGHER OUT OF THE WATER, PILING THE PEOPLE INTO HELPLESS HEAPS AROUND THE STEEP DECKS, AND BY THE SCORE INTO THE ICY WATER."

THEN ONE OF THE SHIP'S FUNNELS FELL.

IT PLUNGED INTO THE SEA . . .

. . . AND MADE A HUGE WAVE THAT THREW THE LIFEBOAT CLEAR OF THE SINKING SHIP.

LIGHTOLLER SCRAMBLED ONTO THE OVERTURNED LIFEBOAT AND LOOKED BACK AT THE *TITANIC*.

"LIGHTS ON BOARD THE *TITANIC* WERE STILL BURNING, AND A WONDERFUL SPECTACLE SHE MADE, STANDING OUT BLACK AND MASSIVE AGAINST THE STARLIT SKY; MYRIADS OF LIGHTS STILL GLEAMING THROUGH THE PORTHOLES, FROM THAT PART OF THE DECKS STILL ABOVE WATER."

2:18 A.M.

THEN THE LIGHTS WENT OUT.

"THE NEXT MOMENT, THE MASSIVE BOILERS LEFT THEIR BEDS AND WENT THUNDERING DOWN WITH A HOLLOW RUMBLING ROAR, THROUGH THE BULKHEADS, CARRYING EVERYTHING WITH THEM THAT STOOD IN THEIR WAY."

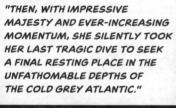

"THEN, WITH IMPRESSIVE MAJESTY AND EVER-INCREASING MOMENTUM, SHE SILENTLY TOOK HER LAST TRAGIC DIVE TO SEEK A FINAL RESTING PLACE IN THE UNFATHOMABLE DEPTHS OF THE COLD GREY ATLANTIC."

"ALMOST LIKE A BENEDICTION EVERYONE ROUND ME ON THE UPTURNED BOAT BREATHED THE TWO WORDS."

SHE'S GONE.

Lightoller's account of the sinking of the *Titanic* is one of many—and there's a lot of variation in how witnesses described the ship's last moments. Lightoller and others said the ship sank in one piece, but others reported that it broke in half.

"AFTER SHE GOT TO A CERTAIN ANGLE, SHE EXPLODED, BROKE IN HALVES, AND IT SEEMED TO ME AS IF ALL THE ENGINES AND EVERYTHING THAT WAS IN THE AFTER PART SLID OUT INTO THE FORWARD PART, AND THE AFTER PART CAME UP RIGHT AGAIN, AND AS SOON AS IT CAME UP RIGHT DOWN IT WENT AGAIN."
—FRANK OSMAN, *TITANIC* CREW

"SHE WENT DOWN AS FAR AS THE AFTER FUNNEL, AND THEN THERE WAS A LITTLE ROAR, AS THOUGH THE ENGINES HAD RUSHED FORWARD, AND SHE SNAPPED IN TWO, AND THE BOW PART WENT DOWN AND THE AFTER PART CAME UP AND STAID UP FIVE MINUTES BEFORE IT WENT DOWN."
—EDWARD JOHN BULEY, *TITANIC* CREW

"TO OUR AMAZEMENT, THE TITANIC BROKE IN HALF— THE PROW SLIPPING DOWN QUIETLY UNDER THE WATER—AND THE STERN REMAINING IN AN UPRIGHT POSITION FOR A COUPLE OF MINUTES, SEEMING TO SAY 'GOODBYE—SO SORRY!'"
—RUTH BECKER, TWELVE-YEAR-OLD SURVIVOR

It would be more than seventy years before researchers would find the wreck of the *Titanic* and learn the truth.

SEVEN

RESCUE AT SEA

No matter how they reported the sinking of the big ship, nearly all the *Titanic*'s survivors shared the same terrible memory: They were haunted by the cries from people in the icy water. First-class passenger Archibald Gracie had still been clinging to the *Titanic*'s rail when the ship went down. He held on and was pulled so far underwater that he felt the pressure in his ears. Once he regained his bearings and swam to the surface, he found some wooden wreckage, perhaps from a crate, and grabbed that to help him stay afloat. He couldn't see the ship anymore at that point, but he heard voices calling out for help.

Gracie said the cries went on for an hour, "but as time went on, growing weaker and weaker until they died out entirely." Eventually, Gracie spotted collapsible lifeboat B overturned in the water, swam to it, and climbed aboard. A dozen other swimmers followed him, but by then, the boat was sitting much lower in the water.

"There were men swimming in the water all about us," Gracie wrote. "One more clambering aboard would have swamped our already crowded craft. The situation was a desperate one, and was only saved by the refusal of the crew, especially those at the stern of the boat, to take board another passenger." Gracie said the men used paddles to steer the boat away from others who were floating on debris in the water.

"Hold on to what you have, old boy," the crew told one man. "One more of you aboard would sink us all."

Gracie said the man replied:

ABOUT THOSE LIFE JACKETS

Most of the people who ended up in the water when the *Titanic* went down were wearing life jackets. The ship had no shortage of those. They were filled with cork, and the crew had even given a safety talk on how to use them.

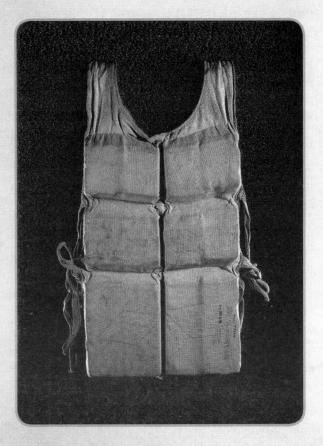

But a life jacket keeping you afloat in icy water doesn't protect you from the cold.

To get some sense for how cold the water of the North Atlantic was that night, fill a big bowl with ice cubes, add water, and swish it around for a minute. Now plunge your hand into the water and hold it there for a few seconds. Painful, isn't it? Imagine trying to swim with your whole body immersed in that painful, frigid water. You wouldn't last long.

The night of the *Titanic* disaster, the water was only twenty-eight degrees. At that temperature, a person would be exhausted or unconscious within just fifteen minutes and dead less than half an hour later. Life jackets simply weren't enough, and many of the *Titanic*'s victims died from the cold.

By now you might be remembering all those lifeboats that launched half-full. Where were they? And how come they weren't picking up some of those people in the water? The answer is that people were afraid. They worried that the frantic, desperate people in the water would grab on to their lifeboat all at once and end up capsizing it.

"We longed to return and pick up some of those swimming," said Ruth Becker, who was in lifeboat 13, "but that should have meant swamping our boat and further loss of lives of us all."

A few lifeboats did go back, rescuing another fourteen people. But the water was frigid and the night was cold, and some of those rescued were already in bad shape. Only half of those saved in lifeboats after the ship went down survived.

Meanwhile, those in the lifeboats waited and waited. When Archibald Gracie realized that wireless operator Harold Bride was on the same lifeboat, he asked which boats he'd been able to contact. Bride told him how the *Baltic,* the *Olympic,* and the *Carpathia* had all responded. The men strained their eyes, searching the horizon. Finally, lights appeared.

2:40 A.M.

ABOUT TWENTY MINUTES AFTER THE SHIP WENT DOWN, THE *CARPATHIA'S* CAPTAIN SAW A GREEN FLARE THAT ONE OF THE *TITANIC'S* OFFICERS SENT UP FROM LIFEBOAT 2.

4:10 A.M.

IT TOOK THE *CARPATHIA* AN HOUR AND A HALF TO GET THERE AFTER THE *TITANIC* WENT DOWN.

THE REST OF THE LIFEBOATS WERE SCATTERED OVER A FOUR- OR FIVE-MILE AREA.

IT WAS DAWN WHEN ARCHIBALD GRACIE AND THE OTHERS ON THE COLLAPSIBLE LIFEBOAT SPOTTED THE RESCUE SHIP.

THERE IS A STEAMER COMING BEHIND US.

THEY COULD SEE LIFEBOATS ROWING TOWARD THE SHIP.

FOUR OF THOSE OTHER LIFEBOATS WERE TIED TOGETHER. THEY HAD ROOM, SO LIGHTOLLER BLEW HIS WHISTLE TO CALL THEM.

COME OVER AND TAKE US OFF!

AYE, AYE, SIR!

TWO OF THE LIFEBOATS CAME OVER AND CAREFULLY, ONE BY ONE, TOOK ON THE MEN WHO'D BEEN BALANCED ON THE COLLAPSIBLE LIFEBOAT ALL NIGHT.

112

FINALLY, THE LIFEBOATS REACHED THE CARPATHIA, AND GRACIE CLIMBED THE ROPE LADDER TO SAFETY.

"I ENTERED THE FIRST HATCHWAY I CAME TO AND FELT LIKE FALLING DOWN ON MY KNEES AND KISSING THE DECK IN GRATITUDE FOR THE PRESERVATION OF MY LIFE."

WHEN LAWRENCE BEESLEY AND THE OTHER MEN IN LIFEBOAT 13 SAW THE LIGHTS OF THE CARPATHIA, THEY STARTED ROWING—AND SINGING.

PULL FOR THE SHORE, SAILOR, PULL FOR THE SHORE . . .

LEAH AKS, THE WOMAN WHO'D BEEN SEPARATED FROM HER BABY, WAS ON LIFEBOAT 13, TOO. ONCE SHE WAS RESCUED, SHE FOUND HIM IN THE ARMS OF ANOTHER PASSENGER WALKING BY.

WHILE MALE SURVIVORS CLIMBED ROPES AND NETS ONTO THE CARPATHIA, WOMEN WERE HAULED UP IN SLINGS AND CHAIRS.

SEVEN-YEAR-OLD EVA HART FROM LIFEBOAT 14 RECALLED BEING "WINCHED UP IN A MAIL BAG BECAUSE CHILDREN COULDN'T CLIMB UP ROPE LADDERS, SO WE WERE EACH ONE OF US PUT IN A SMALL SACK . . ."

". . . AND THAT WAS TERRIFYING, SWINGING ABOUT OVER THE OCEAN."

The *Carpathia* rescued more than seven hundred *Titanic* survivors. Once they were all aboard, the ship turned and began steaming back to New York, where the newspapers were waiting to share the survivors' stories.

Early news coverage of the disaster was spotty. In fact, some of it really couldn't have been farther from the truth.

TITANIC SINKING; NO LIVES LOST, one read.

TITANIC'S PASSENGERS SAVED: LINER BEING TOWED TO HALIFAX, another proclaimed. The mistakes likely happened because there was so much wireless traffic that night, and people got confused. While the *Titanic* was sinking, there was also an oil tanker that was having problems in the Atlantic and being towed to shore. It's possible that people heard about a ship being towed and assumed it was the *Titanic*.

It wasn't. And by the time those papers hit the streets, the ship was at the bottom of the Atlantic. The papers that got it wrong ran corrections later.

Other newspapers waited for facts and got the story right, or at least mostly right. The *New York Times* headline read:

When the *Carpathia* arrived in New York, the big newspapers sent tugboats out to meet it. Reporters tried to talk their way onto the ship to interview survivors, but only two had any luck. The *New York Times* managed to sneak a reporter on board with Guglielmo Marconi, who went to talk with the wireless operators, and they scored an interview with Harold Bride that was published as a big story the next day.

But perhaps the biggest *Titanic* scoop went to a reporter named Carlos Hurd. He was supposed to be on vacation, traveling to the Mediterranean on the *Carpathia* with his wife, Katherine. Before the couple left on their voyage, they had to travel to New York to board the ship. While they were in the city, Hurd visited the offices of the *New York World,* a big paper owned by the Pulitzer family, who also owned the *St. Louis Post-Dispatch,* where Hurd worked. When the *Carpathia*'s voyage was cut short due to the rescue mission, Hurd realized he was in the middle of a huge news story and recruited his wife to help interview *Titanic* survivors.

It was a tricky job because the captain of the *Carpathia* had ordered a news blackout. Passengers weren't allowed to send or receive wireless messages, and the Hurds couldn't even buy paper to write on. So Carlos and his wife had to sneak around interviewing survivors. In an effort to keep the story from getting out, the ship's officers even searched their cabin sometimes. When that happened, Carlos put all his notes and his story on a chair, and his wife sat on them so they couldn't be found. When the Hurds left the cabin, Katherine hid the papers in her underwear.

When the *Carpathia* arrived in New York, one of the tugboats that pulled up alongside it was from the *New York World,* whose editor had remembered that Hurd was on board the rescue ship. Hurd had tied his story up in a cigar box and added corks so it would float if he had to throw it. And that's what he did. It got snagged in a line, but a crew member freed it and tossed it to the editor of the *World,* who rushed back to the office to get the story printed. Finally, the world would hear the story of the *Titanic*'s survivors, how they'd spent a long night in lifeboats, waiting to be rescued after one of the worst maritime disasters in history.

HEROES AND VILLAINS

As newspaper reports about the *Titanic* disaster were published, stories of heroes and villains emerged. Some of those were true. Some were fair characterizations. Others, not so much.

One of the most heroic stories was that Captain Edward Smith rescued a child just before he went down with his ship. One account had him leaping from the bridge with a baby, swimming to a nearby lifeboat (while somehow still holding on to the baby!), and giving the child to someone in the boat before refusing to take a seat himself and swimming away.

If you're thinking that it seems impossible to leap from a ship, surface, and swim to a lifeboat while holding a baby, you're probably right. There's little evidence to support this heroic tale, but it was a story that made people feel better, so many shared it. Shipbuilder Thomas Andrews was also painted as a hero in early news reports, and the story about him throwing deck chairs into the water to help people seems more plausible.

Meanwhile, American newspapers pounced on Bruce Ismay's escape in a lifeboat and wrote stories that painted him as a villain. Many reports portrayed Ismay as a greedy

White Star Line owner who should have gone down with the ship but instead jumped into a lifeboat, leaving women and children on board. One of those stories was published in a newspaper owned by William Randolph Hearst, who was said not to have liked Ismay very much to begin with.

Despite all the stories that showed Ismay as one of the *Titanic*'s villains, the official British inquiry into the disaster concluded that he hadn't really done anything wrong. High Court judge Lord Mersey said Ismay had helped lots of passengers before he got into a lifeboat himself, and when he did, it was because the boat was being lowered and there was nobody else around.

THERE WAS ROOM FOR HIM AND HE JUMPED IN. HAD HE NOT JUMPED IN, HE WOULD MERELY HAVE ADDED ONE MORE LIFE— HIS OWN—TO THE NUMBER OF THOSE LOST.

EIGHT
INQUIRIES AND INVESTIGATIONS

The world was in shock after the *Titanic* went down, and entire communities were grieving. In Southampton, flags were flown at half-mast. Shopkeepers put up shutters and closed for the day. Crowds waited at the White Star Line's offices for news of survivors.

"Southampton is a town of mourning," the *Daily Sketch* newspaper wrote. "Ninety per cent of the crew of the *Titanic* were born or had residences in the port and its neighborhood, and there is scarcely a house but is apprehensive for the safety of a relative or friend."

The city of Belfast, where the *Titanic* was built, was devastated.

When the *Titanic* left Belfast everyone credited a great future for her. There was no vessel afloat on which so much money had been spent, so much human ingenuity exercised, so much care taken to secure the comfort and safety of the passengers. So elaborate were the precautions which had been adopted to guard against accident that it was thought to be practically impossible for her to sink.

—*BELFAST NEWS-LETTER, APRIL 17, 1912*

So what happened? How did the great ship end up sinking? Who was to blame? And how could it have been prevented? Those questions became the focus of two separate inquiries—one on each side of the Atlantic.

In today's world, if there's an airplane crash, officials recover something called the "black box" from the airplane. It's an electronic flight data recorder that saves audio from the cockpit as well as data, such as the plane's altitude and speed. It helps investigators piece together what went wrong.

But when the *Titanic* sank, there was no black box. Investigators had to call witnesses and gather evidence, putting together thousands of pieces of a jigsaw puzzle in order to figure out what happened. Based on witness testimony and records from the ship, they were able to piece together a timeline of the disaster.

TITANIC OFFICER CHARLES LIGHTOLLER WAS ONE OF THE KEY WITNESSES FOR THE BRITISH COMMISSION OF INQUIRY INTO THE LOSS OF THE SS TITANIC, WHICH TOOK PLACE IN MAY AND JUNE 1912.

THOMAS SCANLAN, WHO REPRESENTED THE NATIONAL SAILORS' AND FIREMEN'S UNION, ASKED LIGHTOLLER WHY THE SHIP WAS GOING SO FAST—TRAVELING AT 21.5 KNOTS—EVEN THOUGH THERE WERE KNOWN TO BE ICEBERGS AROUND.

IS IT NOT QUITE CLEAR THAT THE MOST OBVIOUS WAY TO AVOID IT IS BY SLACKENING SPEED?

NOT NECESSARILY THE MOST OBVIOUS.

WELL, IS IT ONE WAY?

IT IS ONE WAY. NATURALLY, IF YOU STOP THE SHIP YOU WILL NOT COLLIDE WITH ANYTHING.

LIGHTOLLER WAS ASKED IF HE FELT PRESSURE TO KEEP GOING SO FAST. HE SAID NO. AND THE CONVERSATION GREW HEATED.

AM I TO UNDERSTAND, EVEN WITH THE KNOWLEDGE YOU HAVE HAD THROUGH COMING THROUGH THIS TITANIC DISASTER,

AT THE PRESENT MOMENT, IF YOU WERE PLACED IN THE SAME CIRCUMSTANCES, YOU WOULD STILL BANG ON AT 21.5 KNOTS AN HOUR?

WHAT I WANT TO SUGGEST TO YOU IS THAT IT WAS RECKLESSNESS, UTTER RECKLESSNESS, IN VIEW OF THE CONDITIONS WHICH YOU HAVE

DESCRIBED AS ABNORMAL, AND IN VIEW OF THE KNOWLEDGE YOU HAD FROM VARIOUS SOURCES THAT ICE WAS IN YOUR IMMEDIATE VICINITY, TO PROCEED AT 21.5 KNOTS?

LIGHTOLLER TOOK ISSUE WITH THE PHRASE "BANG ON" TO SUGGEST HE'D BEEN GOING TOO FAST. BUT SCANLAN DIDN'T BACK OFF.

The inquiry concluded that the collision was, in fact, due to excessive speed and also that a proper watch wasn't kept on the *Titanic*. It said that the lifeboats were properly lowered but not staffed well enough, and that there was no intentional discrimination against third-class passengers, even though fewer of them survived. The investigation also looked at how other ships responded to the *Titanic*'s distress calls and determined that the *Californian* might have made it in time to save more people if it had tried to come sooner.

Officials also made some suggestions for the future—more watertight compartments in ships like the *Titanic,* better rules for lookouts, and enough lifeboats to save everyone on board in the case of an emergency.

Meanwhile, the US Senate conducted its own hearings into the disaster. One question was why people in those partially full lifeboats didn't go back to help.

HAROLD LOWE, A TITANIC CREW MEMBER, TOLD THE SENATE THAT LIFEBOAT 14 HAD FIFTY-EIGHT PEOPLE IN IT WHEN IT REACHED THE WATER.

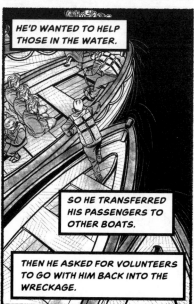

HE'D WANTED TO HELP THOSE IN THE WATER.

SO HE TRANSFERRED HIS PASSENGERS TO OTHER BOATS.

THEN HE ASKED FOR VOLUNTEERS TO GO WITH HIM BACK INTO THE WRECKAGE.

LOWE TOLD THE SENATE THAT HE WAITED FOR AN HOUR AND A HALF— UNTIL THE SCREAMS SUBSIDED— TO GO BACK.

IT WOULD NOT HAVE BEEN WISE OR SAFE FOR ME TO HAVE GONE THERE BEFORE BECAUSE THE WHOLE LOT OF US WOULD HAVE BEEN SWAMPED AND THEN NOBODY WOULD HAVE BEEN SAVED.

127

Testimony at the hearings wasn't limited to the passengers and crew of the *Titanic*. The Senate also had questions about why other ships that were nearby didn't come more quickly.

Ernest Gill, an engine room worker on the *Californian*, testified that he'd seen the lights of a nearby steamer the night the *Titanic* sank. That was at about 11:50 p.m., and his own ship was stopped, surrounded by ice. At around 12:30 a.m., he saw a rocket about ten miles away. He wondered if it was a shooting star, but there was another one ten minutes later. A ship in distress!

So what would you do if you were on a ship and saw another ship sending up distress rockets? You'd probably let somebody know, right? That's not what Gill did.

He went to bed, figuring somebody else would deal with the rockets. The next morning at 6:40 a.m., the *Californian*'s chief engineer woke him up and told him that the *Titanic* had gone down. Other crew members, it turns out, had seen the rockets, too.

"Why the devil didn't they wake the wireless man up?" Gill heard someone ask. He said the entire crew was talking about how the *Californian* pretty much ignored the *Titanic*'s call for help. It was 5:00 a.m.— hours after those rockets went up—when the ship finally started moving.

The sad thing is, the *Californian* was actually closer to the *Titanic* than the *Carpathia* was—only about twenty miles away, compared to fifty-three miles. If the *Californian*'s wireless operator hadn't turned off his equipment . . . if he'd heard the *Titanic*'s distress call . . . and if the *Californian* had taken off on a rescue mission right away, it might have arrived in time to save more people.

Captain Stanley Lord of the *Californian* said that just wasn't true. He said even if he'd known about the *Titanic* sooner, he couldn't have gotten there any faster because it wouldn't have been safe to steam through all that ice in the middle of the night.

At the end of the US hearings, the Senate

officially condemned the *Californian*'s captain for failing to respond to those distress signals and praised Captain Rostron from the *Carpathia* for his rescue. The Senate also suggested changes that might prevent future disaster.

There were big changes in maritime laws and procedures after the *Titanic* sank. Just days later, the routes followed by passenger liners were moved south so they'd be less likely to encounter ice. In 1915, an International Ice Patrol was established that continues—from the sea and the sky—even today. The only times it didn't happen regularly were during the two world wars.

Lifeboat laws changed, too. In 1914, the International Conference for Safety of Life at Sea agreed that every ship should have enough lifeboats for all passengers. The new regulations also required drills and training for crew members.

ABOUT THOSE LIFEBOATS . . .

You'd think that more lifeboats would be a great thing. How could that go wrong? But the new law actually created some new problems. Back in 1914, lifeboats were big, heavy wooden vessels. So while it was great for an ocean liner to have the new legal numbers, it was tricky for small boats that operated in smaller, shallower bodies of water. It was an engineering problem. Too much weight in lifeboats would make the boats top-heavy—and dangerous.

In 1915, a Great Lakes riverboat called the SS *Eastland* rolled over onto its side while it was tied to a dock in Chicago, killing 844 people. The boat had some design issues to begin with and had only been built to handle six lifeboats. With the new law, it was carrying eleven, along with thousands of pounds of life rafts and life jackets that made it unstable and led to the disaster.

NINE
SEARCH FOR THE *TITANIC*

The part of the North Atlantic where the *Titanic* went down is cold, dark, and deep. After the ship sank, no one would see it for more than seventy years. For much of that time, the technology to explore the deep ocean simply didn't exist. Once it did, researchers and explorers began dreaming about how they might be the ones to find the legendary shipwreck.

One of those explorers was a young diver named Robert Ballard. He'd grown up in California, where he loved walking by the ocean as a kid, combing the shoreline to see what washed up on the beach. He'd climb around, checking out marine life in tide pools, and eventually learned to scuba dive.

In 1967, Ballard was assigned to Woods Hole Oceanographic Institution (WHOI) in Massachusetts as a navy oceanographer. While there, he joined a club for divers called the Boston Sea Rovers. He'd hang out with older, more experienced divers who wondered what it would be like to find and swim through the famous ship. Ballard became obsessed with the idea. He wanted to be the one to discover the *Titanic*!

After Ballard left the navy, he stayed at Woods Hole and began working with the *Alvin* Group. *Alvin* was a three-person submarine that had been developed for underwater exploration in 1964 and was updated with new technology in 1973. Workers replaced the old steel hull with one made of a titanium alloy to allow for deeper dives. Now the little submarine could descend to thirteen thousand feet—deep enough to put the wreck of the *Titanic* within reach.

Ballard wrote proposals to raise money for an expedition to find the lost ship. At first, everybody turned him down. But then he talked with Bill Tatum from the *Titanic* Historical Society, who shared his dream. The two of them talked with Alcoa Aluminum and made arrangements to use the company's salvage ship, the *Seaprobe*. But where, exactly, should they begin the search?

The *Titanic*'s official distress call gave the ship's position as 41°46' N, 50°14' W. But officials wondered if that was right, because the *Carpathia* reached the first lifeboats sooner than it should have, if that was really where the ship went down. As a result, experts guessed that the real location of the wreck would be southeast of those coordinates.

Ballard and his crew set out on the *Seaprobe* in October 1977. But within days, the expedition ended in disaster. A drill pipe broke, and with a huge crash, the team lost practically all their expensive setup.

Ballard worried no one would be willing to loan them equipment for another try. What if someone else got to the *Titanic* first? In 1980, Ballard learned that another team of scientists was heading out to search for the shipwreck. But that group didn't have any luck, either.

Meanwhile, Ballard started working with the navy to create a deep-sea imaging system called *Argo*. The navy had its own reasons to be interested in the technology and agreed to work with Ballard to test the system during the summer of 1985. Woods Hole Oceanographic Institution worked with the French Research Institute for Exploitation of the Sea, with Ballard and French researcher Jean-Louis Michel as team leaders. Their high-tech equipment was ready to go. Now where should they search?

THE RESEARCHERS STARTED BY LOOKING AT WHAT THEY ALREADY KNEW ABOUT THE *TITANIC* . . .

. . . LIKE WHERE ITS DISTRESS CALL CAME FROM AND WHERE THE LIFEBOATS WERE FOUND.

THEY NARROWED THE SEARCH TO AN AREA OF ABOUT A HUNDRED SQUARE MILES AND SET OUT ON THE FRENCH VESSEL LE SUROIT IN JULY 1985.

THE SHIP USED SIDE-SCANNING SONAR TO MAP THE OCEAN FLOOR.

BUT STORMY WEATHER MADE THE WORK DIFFICULT.

AFTER THIRTY-ONE DAYS OF SEARCHING WITH NO LUCK, LE SUROIT WAS OUT OF TIME.

AUGUST 15

A FEW WEEKS LATER, BALLARD AND THREE OF THE FRENCH SCIENTISTS SET OUT ON THE KNORR, A SHIP OWNED BY THE NAVY AND OPERATED BY WHOI.

THE EARLIER EXPEDITION HAD ELIMINATED 75 PERCENT OF THE TITANIC SEARCH AREA. AT LEAST NOW SCIENTISTS KNEW WHERE THE SHIP WASN'T.

BY THE TIME THE KNORR GOT TO THE LOCATION, RESEARCHERS HAD JUST TWELVE DAYS TO SEARCH.

THEY SPENT HOUR AFTER HOUR STARING AT SCREENS WITH MILES OF EMPTY OCEAN FLOOR.

SEPT. 1

AS TIME PASSED, THE TEAM WONDERED IF THEY'D HAVE TO RETURN UNSUCCESSFUL ONCE AGAIN.

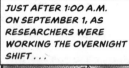

THERE'S SOMETHING.

IT'S COMING IN!

WRECKAGE!

BINGO!

THEY ADJUSTED THE CAMERA ON ARGO, AND IMAGES CAME INTO FOCUS FROM MORE THAN TWELVE THOUSAND FEET BELOW.

WOO-HOO!

ALL RIGHT!

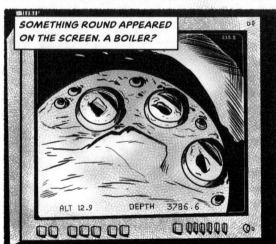

SOMETHING ROUND APPEARED ON THE SCREEN. A BOILER?

ALT 12.9   DEPTH 3786.6

THEY COMPARED THE HAZY IMAGE TO A PICTURE FROM A BOOK THAT SHOWED BOILERS FROM THE TITANIC AND THE OLYMPIC.

NOW THEY WERE SURE. THEY WERE LOOKING AT DEBRIS FROM THE TITANIC.

THE SHIP'S COOK HAD GONE TO FIND BALLARD.

THE GUYS THINK YOU SHOULD COME DOWN. . . .

BALLARD JUMPED OUT OF BED AND PULLED ON HIS JUMPSUIT OVER HIS PAJAMAS.

THE CREW REPLAYED THE VIDEO TO SHOW HIM THE MOMENT ARGO PASSED OVER THE *TITANIC'S* BOILER.

THEY CELEBRATED AS PIECES OF HULL PLATES AND RAILINGS CAME INTO VIEW.

141

2:00 A.M.

THEN SOMEONE NOTICED THE CLOCK, AND THE WEIGHT OF THAT MOMENT SANK IN.

THEY'D DISCOVERED THE *TITANIC* AT ALMOST THE EXACT TIME IT SANK.

BALLARD AND SOME OTHERS WENT OUT TO THE SHIP'S FANTAIL THEN.

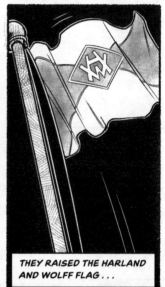

THEY RAISED THE HARLAND AND WOLFF FLAG . . .

. . . AND OBSERVED A FEW MINUTES OF SILENCE.

They'd done it. They'd found the *Titanic*. And now there was more work to do. For the next twenty-four hours, the crew took photos, mapped the ship's debris field, and located the hull.

The crew had a couple more days to take photos, but one day was stormy, and then it was time for the research vessel to return to Woods Hole. When it arrived on September 9, the dock was mobbed with people waiting to see the team that had found the *Titanic* after seventy-three years. The team held a news conference to share details.

The *Knorr* arrives at the WHOI dock
after the *Titanic* discovery.

TOP-SECRET SCIENCE

When Ballard and his fellow researchers first discovered the wreck of the *Titanic*, they didn't share the whole story about how they found it. That's because part of that story was a top-secret military mission! That's why the navy had been willing to fund the technology. It wanted to investigate the wrecks of two American nuclear submarines that sank in the 1960s, the USS *Thresher* and the USS *Scorpion*.

The navy wanted to know what was happening with the nuclear reactors that powered the two subs. Was it safe for them to be underwater like that, or were they deteriorating down there? Military officials also wanted to know if there was any evidence to suggest that the *Scorpion* might have been shot down by the Soviets.

The USS *Scorpion* off the coast
of Connecticut in 1960

The navy sent Ballard out to survey the two wrecked submarines and promised he could use whatever time was left to look for the *Titanic*. The *Titanic* wreck was located in between those two submarines, so it worked out pretty well for everyone. Ballard used the equipment to get photos of the two submarines. He reported back to the navy that they were safe at the bottom of the ocean and not harming the environment, as far as he could tell. It looked like a pipe failure caused the *Thresher* to sink, and while it wasn't clear what had happened to the *Scorpion,* there was no evidence that it had been shot down. Once that work was done, Ballard's team was left with just twelve days to search for the *Titanic.* It turned out to be just enough time to find it.

TEN
MORE RESEARCH, MORE ANSWERS

The discovery of the *Titanic* in 1985 was only the beginning. Ballard returned to the *Titanic* the following year, and this time, the ship was much easier to find. Researchers already knew its exact location, 380 miles southeast of Newfoundland, which was thirteen miles southeast of the position given in its final distress call. On this expedition, the team was able to take a lot more photos. More details took shape about the ship's last moments, too. Researchers confirmed that it did, in fact, break in half before it sank.

Scientists who have been studying the ship and its components also have some new ideas about *why* the *Titanic* sank. Though much of the hull is buried in mud, experts were able to use sonar to map the damage to those six watertight compartments. Fragments of steel were also recovered from the hull, and when scientists studied it, they noticed two big problems. Researchers from Canada's Defence Research Establishment Atlantic (DREA) found that the steel had jagged edges, suggesting something called brittle fracture, which happens when metals that are supposed to be pliable become more rigid. That can be caused by low temperatures, high sulfur content in steel, and high-impact speeds—all three of which came into play the night the *Titanic* hit the iceberg. The science behind this wasn't known at the time the *Titanic* was built. The ship's engineers didn't know about brittle fracture, so they couldn't have predicted what would happen.

MEET THE ROBOT RESEARCHERS

The human explorers and researchers who located the wreck of the *Titanic* and studied it later on had help from a team of robots. Woods Hole Oceanographic Institution has a small fleet of robotic researchers that aid scientists in exploring the deep sea.

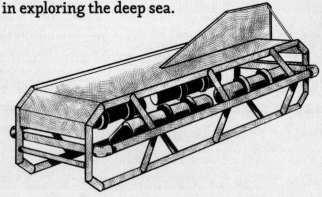

Argo is the system of video cameras and sonar that helped scientists locate and see the wreck of the *Titanic*. It's towed by a surface vessel—the *Knorr*, in the case of the *Titanic* discovery—and can take wide-angle video from fifty to one hundred feet above the ocean floor. Systems like *Argo* often explore an undersea location before humans go down to check it out.

When researchers returned to the *Titanic* in 1986, they had a new ship and some new vehicles to work with.

Alvin is a three-person submarine, also known as a DSV, or deep-submergence vehicle. It can take scientists fourteen thousand feet under the sea in a dive that lasts up to eight hours. It takes two hours to go down and another two to get back to the surface, so researchers have about four hours to take photographs, collect samples, and do experiments. *Alvin* has made almost five thousand

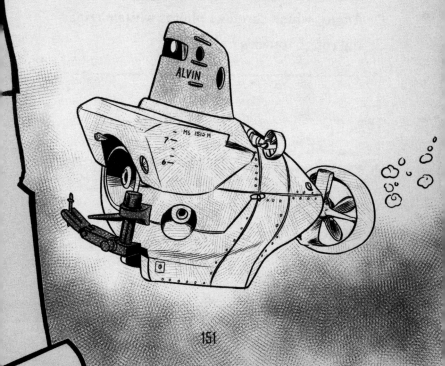

dives, including those that helped map the wreck of the *Titanic*.

Jason Jr. is a remotely operated vehicle, or ROV, that helped scientists on their 1986 *Titanic* mission. It's attached to *Alvin* by a fiber-optic cable that's three hundred feet long and is controlled by a pilot inside *Alvin*. Because *Jason Jr.* is smaller, it can go places that *Alvin* can't fit. When it was surveying the *Titanic*, *Jason Jr.* towed an imaging sled called *Angus*, which can take photos while navigating rough underwater terrain.

You might think that once the *Titanic* had been found, people would agree to leave it alone. It is a grave site, after all, right? But there have been numerous expeditions since then, including some that salvaged artifacts from the wreck and others that even brought tourists to see it.

Since 1987, a private company called RMS Titanic Inc. has brought up around 5,500 artifacts from the wreck, including a piece of the hull, china from the dining rooms, and passengers' personal possessions.

The artifacts were displayed at special exhibitions where people could pay to see them until the company that ran those displays went out of business in 2016. Several museums got together to try to return the artifacts to the United Kingdom, but a group of investors ended up getting them instead.

In 1995, filmmaker James Cameron visited the shipwreck. He hired two Russian submarines, built special cameras, and put them in titanium cases that could withstand the pressure of the ocean at that depth. He ended up using the footage in *Titanic,* his famous movie about the disaster.

In the 1990s, you didn't have to be a famous movie director to visit the ship. You just had to have a lot of money. Several groups started offering tourists the chance to go down in a submersible to see the *Titanic* firsthand. One couple even got married in a submarine on the deck of the ship in 2001.

The last of those controversial trips ended in 2012, but another private company has announced plans to start them up again soon. It'll cost over $100,000 per person to go.

All this time, there was more research happening on the shipwreck, too. Historians still had so

many questions about how the *Titanic* sank. In 2010, they got some answers when autonomous robots went down to map the site. They took more than a hundred thousand photographs, and the team also used side-scanning sonar to make a complete map of the wreck and its debris field. It showed how hundreds of objects and ship parts landed on the ocean floor and provided answers to some of those questions. Did the stern plunge straight down or rotate as it sank? Marks on the ocean floor suggest it spun around on its way down.

Other evidence, presented to WHOI by History Channel divers in 2005, suggests that those eye-witness accounts about the ship breaking in half might not be the whole story. Some researchers now wonder if the ship may have broken into three pieces as it sank, which means it could have gone down even more quickly than previously thought. Other scientists disagree with that theory. The more researchers study the *Titanic*, the more they learn. And then even more questions arise. But one thing scientists and historians agree on is the need to protect this historic shipwreck.

2001 CONVENTION ON THE PROTECTION OF THE UNDERWATER CULTURAL HERITAGE

In 2012, the *Titanic* wreck got some special protection from UNESCO, the United Nations Educational, Scientific and Cultural Organization. That group's 2001 Convention on the Protection of the Underwater Cultural Heritage was established to protect sites like the *Titanic* wreck from treasure hunting and overexploration. The sites had to have been underwater for at least a hundred years, though, to qualify for that protection. *Titanic* hit that mark in 2012 and got the official designation then.

Certain kinds of exploration are still allowed, however. In August 2019, researchers made the first crewed dives on the *Titanic* since 2005. A group called Caladan Oceanic made five manned submersible dives on the ship over eight days. They filmed their work for a National Geographic documentary called *Mission Titanic,* and took photographs of the ship that will be used to make 3-D models.

When researchers surfaced from their dives, they had troubling news to share. The *Titanic* is beginning to fall apart. Ocean currents, corrosion from salt, and bacteria that live in the sea have been taking a toll on the famous ship for more than a hundred years now. The hull is starting to collapse. The worst decay is on the starboard side of the officers' quarters. Even though the cold water of the North Atlantic has slowed the ship's decay, mollusks have eaten lots of the wood, and bacteria have been eating the metal, creating what scientists call rusticles—dripping rust formations that look like icicles under the sea.

Scientists are concerned that it could be just a few decades before the *Titanic* deteriorates completely. While most people won't have the chance to visit before then, there are other ways to learn about the famous ship and imagine what it might have been like to be a passenger.

In 2012, *Titanic* Belfast opened in Northern Ireland. The museum includes documents and artifacts from the ship as well as a ride that takes visitors on a trip through the sights and sounds of the shipyard.

And there may one day be a full-scale replica of the *Titanic* as well. An Australian billionaire has been building one in China since 2012 and has plans to set sail with his *Titanic II* when it is complete. The plan is to go from China to Dubai, head for Southampton to re-create the original ship's voyage (hopefully without the iceberg this time), and continue around the globe.

The multimillion-dollar replica will have the same cabin layout and dining rooms as the original but better safety and navigation features. Even so,

A 3-D rendering of the *Titanic II*

not everyone approves of the project. Some say it's just not cool to re-create a ship that ended up being the final resting place for so many victims of the wreck.

More than a century after the disaster, the story of the *Titanic* continues to captivate the world—and there's been no end to the legends and rumors surrounding the shipwreck. Some of them are true—or partly true, anyway. The rest? Well, they need to be smashed with some facts.

TITANIC
RUMORS

Are they true?
Partly true?
Or smash-worthy?

THE RUMOR:
THE *TITANIC* NEVER REALLY SANK!

THE DETAILS: Some conspiracy theorists claim that it was actually the *Titanic*'s sister ship, the *Olympic,* that went down. They've suggested that the White Star Line switched the two ships before the *Titanic*'s voyage in a scheme designed to collect insurance money later on. The theory suggests that after the *Olympic* had that accident that delayed the launch of the *Titanic,* it was too damaged to be profitable. So maybe the owners sent out the *Olympic* in place of the *Titanic,* knowing it would likely sink and they'd get to collect the insurance money!

THE REAL DEAL: There's absolutely no evidence to support this theory. The two ships were similar, but they weren't identical, and it would have been almost impossible for the White Star Line to pass off a year-old ship as brand-new. This story needs smashing!

THE RUMOR:
A COAL FIRE DOOMED THE *TITANIC* BEFORE ITS MAIDEN VOYAGE EVEN BEGAN.

THE DETAILS: In a documentary that aired on the Smithsonian Channel in 2017, journalist Senan Molony made the case that a coal fire on the *Titanic* weakened the ship's hull, making it more vulnerable to damage from the iceberg.

THE REAL DEAL: There are photographs from the shipyard that show a black streak on the outside of the hull's starboard side. Engineers say that could have been caused by a coal fire, and there are mentions of such a fire in historical documents. But did the fire cause the disaster? Probably not. Many experts believe that it may have sped things up a little but that the ship would have gone down anyway.

THE RUMOR:
PRICELESS TREASURES SANK WITH THE *TITANIC*!

THE DETAILS: There were rumors after the *Titanic* sank that a famous gem called the Hope Diamond had been on board—and that it was cursed, so that caused the wreck! There were whispers that the diamond was secretly being taken to America, and even though that was never confirmed, at least one newspaper reported it anyway.

THE REAL DEAL: The Hope Diamond wasn't on the *Titanic,* and it's not at the bottom of the ocean now. It's at the Smithsonian Museum of Natural History in Washington, DC, where it doesn't seem to be cursing anyone. But there *were* treasures on the *Titanic*. A number of the ship's wealthy passengers had jewelry with them, and some of that was recovered from the wreck. Other artifacts from the ship, including the bandmaster's violin, have sold at auction for more than a million dollars apiece.

THE RUMOR:
A CURSED MUMMY ON BOARD THE *TITANIC* CAUSED THE WRECK!

THE DETAILS: Okay, so maybe the Hope Diamond didn't curse the ship. But how about a bad-luck mummy? Another rumor that spread after the shipwreck claimed that the supposedly cursed mummy of the princess of Amen-Ra had been purchased by an American collector and was in the hold of the doomed ship.

THE REAL DEAL: A quick look at the ship's manifest, which lists all the cargo on board, shows there were no mummies on the *Titanic*. A passenger on the ship was reportedly talking about the cursed mummy at dinner one night, though, which may be how the rumor started, even though he never claimed it was on the ship.

A *TITANIC* TIMELINE

MARCH 31, 1909—
Construction of the *Titanic* begins as the keel is laid at Harland and Wolff shipyard in Belfast.

MAY 31, 1911—The *Titanic* is launched! The ship then goes to a dry dock for more work.

1912

APRIL 2—The *Titanic* undergoes sea trials and departs Belfast for Southampton.

APRIL 4—The *Titanic* arrives in Southampton, where passengers, crew, and supplies are loaded on board.

APRIL 10—The *Titanic* leaves Southampton and goes to Cherbourg, France, to pick up more passengers. Then the ship departs for Queenstown, in Ireland.

APRIL 11—The *Titanic* picks up more passengers in Queenstown before setting out for New York.

APRIL 14—The *Titanic* collides with an iceberg at 11:40 p.m.

APRIL 15—The *Titanic* sinks at 2:20 a.m. The *Carpathia* arrives to rescue survivors from lifeboats at about 4:00 a.m.

APRIL 18—The *Carpathia* arrives at Pier 54 in New York City with the *Titanic*'s survivors.

1985—An American-French team of researchers locates the *Titanic* and returns to Woods Hole Oceanographic Institution with photographs of the wreck site.

1986—The submersible vehicle *Alvin* explores the wreck, taking more images.

1987—The first salvage expedition on the *Titanic* results in about 1,800 artifacts being collected from the wreck.

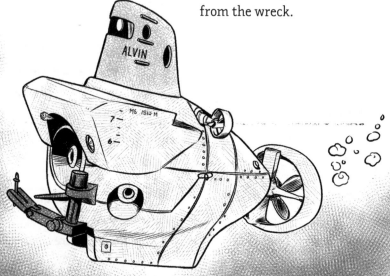

1995—Filmmaker James Cameron visits the wreck to take video footage to use in his movie *Titanic*.

1998—Tourists dive on the wreck of the *Titanic* for the first time.

More artifacts are raised from the ship, including a piece of the hull.

2005—Two crewed submersibles dive to the wreck, bringing back more photographs and video.

2010—The *Titanic* wreck site is mapped by underwater robots using side-scan sonar and more than a hundred thousand photographs.

2012—The *Titanic* is given official protection under UNESCO's 2001 Convention on the Protection of the Underwater Cultural Heritage.

2019—A new expedition to the *Titanic,* the first crewed visit to the shipwreck since 2005, discovers that the ship is deteriorating quickly due to ocean currents, salt, and metal-eating bacteria.

AUTHOR'S NOTE

Like many kids, I was fascinated by the story of the *Titanic* as I was growing up. But mostly I'd heard the romantic legends of the shipwreck instead of the more interesting, true details. And, of course, I'd seen some movies about the famous disaster, which often make for great storytelling but not-so-accurate history.

It was only when I was a grown-up writer doing research that I began to learn the real deal behind the famous disaster. I read pages and pages of historical documents, including news reports and testimony from passengers and crew members who survived. In some cases, these sources contradict one another in the exact numbers and details surrounding the disaster, so this book includes approximations from the most reliable sources. I also had the chance to explore *Titanic* Belfast, an amazing museum built on the site of the old Harland and Wolff shipyard in Belfast, Northern Ireland. I'm most grateful to the staff who answered my many questions about not only the shipwreck itself but the history of the shipyard and the process of building the *Titanic* and the *Olympic*.

If you're interested in the *Titanic* and ever have an opportunity to visit the Belfast museum, you should jump at the chance. I also recommend the following books and websites for readers who would like to know more about the *Titanic* disaster, its discovery, and the later expeditions to locate the wreck.

BOOKS

882 ½ Amazing Answers to Your Questions About the Titanic by Hugh Brewster and Laurie Coulter (Firefly Books, 2018)

Discovering Titanic: *The Story of the World's Most Famous Shipwreck* by Ben Hubbard (Carlton Kids, 2018)

Flying Deep: Climb Inside Deep-Sea Submersible Alvin by Michelle Cusolito, illustrated by Nicole Wong (Charlesbridge, 2018)

If You Were a Kid Aboard the Titanic by Josh Gregory, illustrated by Sebastià Serra (Scholastic, 2017)

Inside the Titanic by Hugh Brewster, illustrated by Ken Marschall (Little, Brown, 1997)

Titanic (National Geographic Readers) by Melissa Stewart (National Geographic, 2012)

Titanic: *Voices from the Disaster* by Deborah Hopkinson (Scholastic, 2014)

Titanic: *Young Survivors* (10 True Tales) by Allan Zullo (Scholastic, 2015)

WEBSITES

"Remembering the *Titanic*" from National Geographic tells the story of the *Titanic*'s voyage and wreck, along with a slideshow of historical images: kids.nationalgeographic. com/explore/history/a-titanic-anniversary

The *Titanic* Belfast website has information about the museum in Northern Ireland as well as online resources: titanicbelfast.com

"Twenty Top *Titanic* Facts" from National Geographic Kids shares some more surprising details about the famous ocean disaster: natgeokids.com/nz/discover/history /general-history/would-you-have-survived-the-titanic

BIBLIOGRAPHY

Ballard, Robert D. *The Discovery of the* Titanic: *How the Greatest Ship Ever Lost Was Found.* New York: Madison Publishing, 1987.

Barron, James. "After Ship Sank, Fierce Fight to Get Story." *New York Times.* April 9, 2012. cityroom.blogs.nytimes .com/2012/04/09/after-the-ship-went-down -scrambling-to-get-the-story.

Behe, George. Titanic: *Safety, Speed and Sacrifice.* Polo, IL: Transportation Trails, 1997.

Berger, Meyer. *The Story of the* New York Times: *1851 – 1951.* New York: Simon & Schuster, 1951.

Blair, William. Titanic: *Behind the Legend.* Belfast: National Museums Northern Ireland, 2012.

Brown, David G. *The Last Log of the* Titanic. Camden, ME: International Marine/McGraw-Hill, 2001.

Bryceson, David. *The* Titanic *Disaster as Reported in the British National Press April–July 1912.* New York: W. W. Norton, 1997.

Cameron, Stephen. Titanic: *Belfast's Own.* Newtownards, Northern Ireland: Colourpoint Books, 2011.

Chapin, Charles E. *Charles Chapin's Story, Written in Sing Sing Prison.* New York: G. P. Putnam's Sons, 1920.

Davie, Michael. Titanic: *The Death and Life of a Legend.* New York: Alfred A. Knopf, 1987.

Eaton, John P., and Charles A. Haas. Titanic: *Destination Disaster: The Legends and the Reality.* New York: W. W. Norton, 1987.

Elias, Leila Salloum. "Alien Passengers: Syrians Aboard the *Titanic.*" The Institute for Palestine Studies. *Jerusalem Quarterly* 52 (2003): 51–67. Accessed online: palestine -studies.org/sites/default/files/jq-articles /JQ-52-Salloum-Alien_Passengers_2.pdf.

Foster, John Wilson, ed. Titanic *Reader.* New York: Penguin, 2000.

Gainey, Caitlin. "*Titanic:* The Reboot." *Scientific American.* August 30, 2019. blogs.scientificamerican.com /observations/titanic-the-reboot.

Gracie, Archibald. *The Truth About the* Titanic. Riverside, CT: 7 C's Press, 1973.

Hughes, Zondra. "What Happened to the Only Black Family on the *Titanic.*" *Ebony.* June 2000.

Hutchings, David F., and Richard de Kerbrech. *RMS* Titanic *Owners' Workshop Manual: An Insight into the Design, Construction, and Operation of the Most Famous Passenger Ship of All Time.* Sparkford, Yeovil Somerset, UK: Haynes Publishing, 2011.

Hyslop, Donald, Alastair Forsyth, and Sheila Jemima. Titanic *Voices: Memories from the Fateful Voyage.* New York: St. Martin's Press, 1994.

Lewis, Danny. "A Coal Fire May Have Helped Sink the *Titanic.*" *Smithsonian Magazine.* January 5, 2017. smithsonianmag.com/smart-news/coal-fire-may-have -helped-sink-titanic-180961699.

Lightoller, C. H. *Titanic.* Riverside, CT: 7 C's Press, 1975.

Lord, Walter. *A Night to Remember: The Classic Account of the Final Hours of the* Titanic. New York: St. Martin's Griffin, 1955.

Loss of the Steamship "Titanic," Report of a Formal Investigation into the Circumstances Attending the Foundering on April 15, 1912, of the British Steamship "Titanic," of Liverpool, After Striking Ice in or Near Latitude 41°46' N., Longitude 50°14' W., North Atlantic Ocean, as Conducted by the British Government. 62nd Congress, 2nd session. Senate document No. 933. Presented by Mr. Smith of Michigan. August 20, 1912. Washington: 1912.

Maxtone-Graham, John. Titanic *Tragedy: A New Look at the Lost Liner.* New York: W. W. Norton & Company, 2011.

Mayo, Jonathan. Titanic: *Minute by Minute.* London: Short Books, 2016.

McCaughan, Michael. *The Birth of the* Titanic. Montreal: McGill—Queen's University Press, 1998.

Moselle, Rebecca. "*Titanic* Sub Dive Reveals Parts Are Being Lost to Sea." *BBC News.* August 21, 2019. bbc.com/news /science-environment-49420935.

National Geographic Staff. "How the *Titanic* Was Lost and Found." *National Geographic.* August 22, 2019. nationalgeographic.com/culture/topics/reference /titanic-lost-found.

O'Donnell, E. E. *Father Browne's* Titanic *Album: A Passenger's Photographs and Personal Memoir.* Dublin: Messenger Publications, 2011.

Parisi, Paula. "Lunch on the Deck of the *Titanic.*" *Wired.*
February 1, 1998. wired.com/1998/02/cameron-3.

Roach, John. "*Titanic* Was Found During Secret Cold War
Navy Mission." *National Geographic.* November 21, 2017.
nationalgeographic.com/news/2017/11/titanic-nuclear
-submarine-scorpion-thresher-ballard.

Rosenberg, Eli. "A Magnet for *Titanic* Fans in an Unexpected
Graveyard." *New York Times.* June 16, 2016.
nytimes.com/2016/06/18/world/what-in-the-world/a
-magnet-for-titanic-devotees-in-a-halifax-graveyard
.html?rref=collection%2Ftimestopic%2FTitanic.

Stranahan, Susan Q. "The Eastland Disaster Killed More
Passengers Than the *Titanic* and the *Lusitania.* Why Has
It Been Forgotten?" *Smithsonian Magazine.* October 27,
2014. smithsonianmag.com/history/eastland-disaster
-killed-more-passengers-titanic-and-lusitania-why
-has-it-been-forgotten-180953146.

Turner, Steve. *The Band That Played On: The Extraordinary
Story of the 8 Musicians Who Went Down with the
Titanic.* Nashville: Thomas Nelson, 2011.

US Senate. Titanic *Disaster: Hearings Before a Subcommittee
of the Committee on Commerce, United States Senate,
Second Session, Pursuant to S. Res. 283, Directing the
Committee on Commerce to Investigate the Causes
Leading to the Wreck of the White Star Liner "Titanic."*
Washington, DC: Government Printing Office, 1912.

Wade, Wyn Craig. *The* Titanic: *End of a Dream.* New York: Rawson, Wade Publishers, 1979.

White, John D. T. *The RMS* Titanic *Miscellany.* Dublin: Irish Academic Press, 2011.

Wilson, Andrew. *Shadow of the* Titanic: *The Extraordinary Stories of Those Who Survived.* New York: Atria Paperback, 2011.

IMAGE CREDITS

George Grantham Bain/Library of Congress Prints and Photographs Division (pp. 37, 39); Barclay Bros. (p. 44 top); Board of Trade/The National Archives (pp. 38 top, 41 top); Brandon Brewer/United States Coast Guard (p. 131); Division of Work and Industry, National Museum of American History, Smithsonian Institution (p. 109); Roderick Eime (p. 161); Everett Collection/Shutterstock.com (p. 29); Harris & Ewing Collection (pp. 40 bottom, 49); Lori Johnston/US National Oceanic and Atmospheric Administration (p. 158); © Nico Kaiser (p. 159); Library of Congress Prints and Photographs Division (p. 47); Max Rigot Selling Company (p. 133); National Oceanic and Atmospheric Administration (p. 143); National Portrait Gallery (p. 42); *The New York Times* (p. 48); Unknown Author/Wikimedia Commons (pp. 2, 26, 36, 38 bottom, 40 top, 41 bottom, 43 top and bottom,

INDEX

SMASH MORE STORIES!

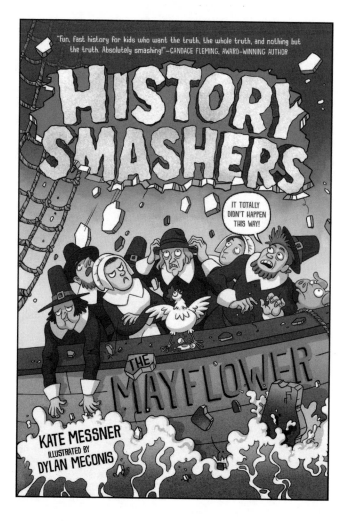

"Fun, fast history for kids who want the truth, the whole truth, and nothing but the truth. Absolutely smashing!" —CANDACE FLEMING, AWARD-WINNING AUTHOR

HISTORY SMASHERS

IT TOTALLY DIDN'T HAPPEN THIS WAY!

THE MAYFLOWER

KATE MESSNER
ILLUSTRATED BY
DYLAN MECONIS

Turn the page for a peek at the first book
in the History Smashers series!

You've probably heard about the *Mayflower*. Chances are, someone told you about the Pilgrims, who came to America because they wanted religious freedom. You probably learned how they crossed the wild Atlantic, how they landed at Plymouth Rock in Massachusetts, how the Wampanoag people taught them to grow corn, and how they all celebrated by sitting down together for a feast—the very first Thanksgiving. But only parts of that story are true.

There's a lot more to the history of the *Mayflower*, the Pilgrims, and the Wampanoag. So let's take a look at the historical documents, smash some of those old myths, and uncover the *real* story.

ONE
WHO WERE THE PILGRIMS, ANYWAY?

f April showers bring May flowers, what do May flowers bring?

The answer to the riddle, of course, is Pilgrims. The joke works because almost everyone knows a little about the Pilgrims. We've heard how they left England

and came to America in search of religious freedom. But that's not even close to the whole story. For starters, the Pilgrims didn't go to America when they left England. Not at first, anyway.

The real-deal story of the *Mayflower* begins way back in the 1530s, when King Henry VIII made some big changes to religion in England. King Henry wanted a son who could grow up to be the king of England, too. He and his first wife only had a daughter, though. Henry decided the solution was to get divorced and marry someone else, with whom he might have a son.

But the Roman Catholic Church was the official church in England then, and it did not allow divorce. King Henry went all the way to the pope, the leader of the whole Catholic Church, to argue that he should be able to leave his wife and marry a new one. When the pope said no, Henry decided to break away from the Catholic Church and start his own. From then on, the Church of England would be the official church of the land.

King Henry wasn't the only one who had issues with the Roman Catholic Church at that time. Many complained that Catholic leaders had too much power and wealth. But not everyone liked King Henry's new church, either. Some thought it was too similar to the Catholic Church. One group, called the Puritans, wanted the new church to be "purified" of all the old practices. Other people didn't think that was enough. They were called Separatists because they wanted to separate from the Church of England completely and have their own religion. The Separatists thought that true Christian believers should come together in their own small churches. They wanted those churches to be independent so members could study the Bible and make decisions on their own.

William Brewster, who was the postmaster of a village called Scrooby, decided to start a church in his own house. It was a risky idea. Back then, people who didn't follow the Church of England could be thrown in jail. In his book *Of Plymouth Plantation*, Pilgrim William Bradford wrote that Brewster's Separatists were "hunted and persecuted on every side."

Government officials were watching the Separatists' houses day and night. Some of them did get thrown in jail. You can probably understand why leaving England was starting to seem like a good idea.

So that's when the Separatists set sail for America, right?

Wrong. They went to Holland.

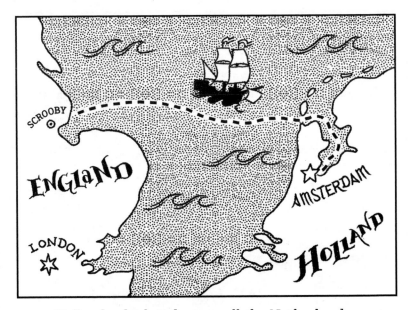

Holland, which today we call the Netherlands, was known for religious freedom. Brewster learned that a small group of Separatists had recently escaped to the city of Amsterdam, where they could practice their religion in peace. That seemed like a good idea, so Brewster made plans to take his group there, too. His followers were nervous, though. They didn't speak Dutch. They weren't sure how they'd earn money to support their families. Bradford later wrote that to many of the Separatists, taking off for Holland seemed like "an adventure almost desperate" and "a misery worse than death." But after much discussion, they decided to go anyway.

1607: BREWSTER ARRANGED FOR A SHIP TO SNEAK HIS CONGREGATION AWAY TO AMSTERDAM. IT WAS EXPENSIVE, AND THEY HAD TO WAIT A LONG TIME, BUT HE DIDN'T SEE ANY OTHER OPTION.

FINALLY THE DAY ARRIVED.

IT WAS TIME FOR THE SEPARATISTS TO LEAVE ENGLAND ONCE AND FOR ALL!

BUT THEN EVERYTHING FELL APART.

THE SHIP'S CAPTAIN HAD RATTED THEM OUT!

THE SHIP'S CREW RANSACKED ALL THE PASSENGERS' BELONGINGS,

LOOKING FOR MONEY.

THEY TURNED THE SEPARATISTS OVER TO THE AUTHORITIES.

INSTEAD OF ESCAPING TO HOLLAND,

THEY ENDED UP SPENDING A MONTH IN AN ENGLISH PRISON.

SMASH EVEN MORE STORIES!

Turn the page for a peek at the third book
in the History Smashers series!

You've probably heard the phrase "Remember Pearl Harbor." The Japanese attack on Pearl Harbor, Hawaii, is one of the most infamous events in American history. Maybe you've heard stories about that day—how the attack was a total surprise, with no warning whatsoever. How it lit up the morning sky on December 7, 1941, devastated the American fleet in the Pacific, and created immediate support for the United States to jump into World War II. But only parts of that story are true. When we look closely at documents from that time period, other parts come crashing down. Here's the real deal about the not-such-a-total-surprise attack that eventually led the United States into its second world war—along with some important but not-so-well-known stories about what happened next.

ONE
WAR
OVERSEAS

People often tell the story of Pearl Harbor as if the Japanese attack in Hawaii happened completely out of the blue, with no warning and nothing to suggest there might be trouble. In reality, there were plenty of warning signs. But to understand how the attack on Pearl Harbor happened, we need to go back in time a bit.

Sometimes Americans think of 1941 as the beginning of World War II, but that was just when the United States got involved. The war had actually started two years earlier, in September 1939. That's when Nazi

Germany invaded Poland. France and Great Britain declared war on Germany two days later.

German troops march into Poland in September 1939.

It wasn't long before the war took over Europe. Between April and June 1940, Germany invaded Denmark, Norway, Luxembourg, the Netherlands, Belgium, and France. In June, Germany picked up some help from Italy, which declared war on Britain and France. And in July, Germany started bombing Great Britain.

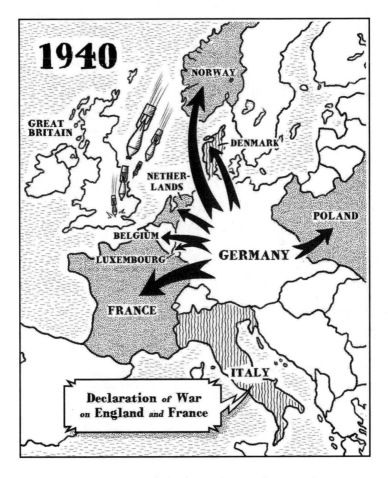

1940

GREAT BRITAIN

NORWAY

DENMARK

NETHER-LANDS

BELGIUM

LUXEMBOURG

GERMANY

POLAND

FRANCE

ITALY

Declaration of War on England and France

Sounds complicated, doesn't it? That's what most Americans thought, too, back when all this was happening. At the time, many people in the United States were in favor of a policy called isolationism, which basically means minding your own business.

NO FOREIGN ENTANGLEMENTS

STAY OUT OF EUROPE'S WAR

President Franklin Delano Roosevelt was concerned about the situation, but he understood that most Americans didn't want to go to war. The US government sent supplies to Great Britain but refused to send troops, even though Britain's prime minister, Winston Churchill, had asked for help. Roosevelt had made a promise to the American people.

WE WILL NOT PARTICIPATE IN FOREIGN WARS.

AND WE WILL NOT SEND OUR ARMY, NAVAL OR AIR FORCES TO FIGHT IN FOREIGN LANDS OUTSIDE OF THE AMERICAS.

EXCEPT IN CASE OF ATTACK.

But that promise got harder and harder to keep. Germany wasn't the only country trying to take over other nations. Japan was doing the same thing.

For two centuries, Japan had kept to itself, an isolated island nation. That started to change in 1853, when Commodore Matthew Perry led a US Navy expedition into what is now called Tokyo Bay, trying to open up Japan to trade with other nations.

brought in to build a naval air force. Then Japan's military leaders decided it was time to expand the empire by taking more land.

In 1931, Japan invaded Manchuria, an area in northeast China. By 1937, it had launched a full-scale invasion of China. Hundreds of thousands of Chinese people were killed. The United States sent millions of dollars in relief funds to China but refused to get involved in the fighting.

When Germany started invading all those other countries in Europe, Japan's military leaders decided to grab even more territory. Since much of Europe was busy fighting the Nazis, it seemed like a perfect time for Japan to invade French Indochina, the part of East Asia that now includes Vietnam, Laos, and Cambodia. American leaders worried the Philippines, then under US control, might be next.

COLONIALISM

Right about now, you might be thinking, "Hey, wait a minute! What are countries like France and the United States doing in East Asia anyway?" The answer to that question has to do with colonialism. That's the policy of taking over other countries, sending settlers, and using those countries' natural resources.

European nations, and later the United States, engaged in this practice throughout history. That's why many people of East Asia weren't in charge of their own lands back then. The United States had taken control of the Philippines from Spain in a war in 1898. And why did Spain have the Philippines to begin with? Because explorer Ferdinand Magellan had landed there in 1521 so Spain could take it from the people who lived there.

I CLAIM THESE ISLANDS FOR THE KING OF SPAIN!

COMING SOON!

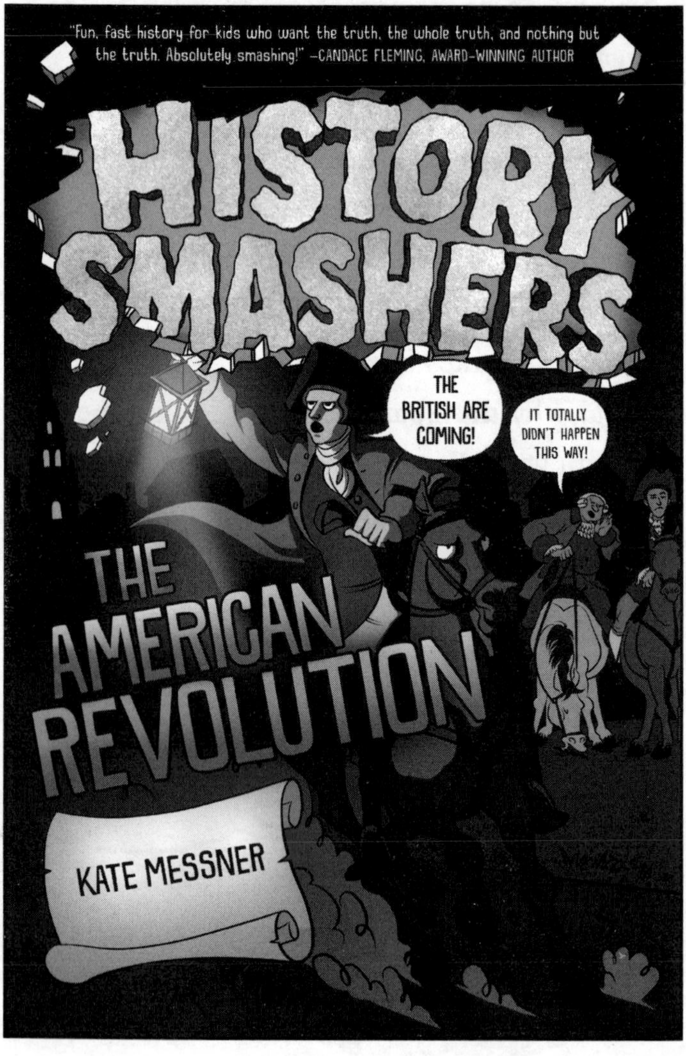
"Fun, fast history for kids who want the truth, the whole truth, and nothing but the truth. Absolutely smashing!" —CANDACE FLEMING, AWARD-WINNING AUTHOR
HISTORY SMASHERS
THE BRITISH ARE COMING!
IT TOTALLY DIDN'T HAPPEN THIS WAY!
THE AMERICAN REVOLUTION
KATE MESSNER